Schriftenreihe aus dem Institut für Strömungsmechanik

Herausgeber
J. Fröhlich, S. Odenbach, K. Vogeler

Institut für Strömungsmechanik
Technische Universität Dresden
D-01062 Dresden

Band 4

Martin Kunze

Experimental Investigations of Platform Film-Cooling in Highly-Loaded Turbine Vanes

TUDpress
2013

Die vorliegende Arbeit wurde am 16. Mai 2012 an der Fakultät Maschinenwesen der Technischen Universität Dresden als Dissertation eingereicht und am 30. September 2013 erfolgreich verteidigt.

This work was submitted as a PhD thesis to the Faculty of Mechanical Science and Engineering of TU Dresden on 16 May 2012 and successfully defended on 30 September 2013.

Gutachter | Reviewers
Prof. Dr.-Ing. Konrad Vogeler, TU Dresden
Prof. Dr.-Ing. habil. Bernhard Weigand, Universität Stuttgart

Bibliografische Information der Deutschen Nationalbibliothek
Die Deutsche Nationalbibliothek verzeichnet diese Publikation in der Deutschen Nationalbibliografie; detaillierte bibliografische Daten sind im Internet über http://dnb.d-nb.de abrufbar.

Bibliographic information published by the Deutsche Nationalbibliothek
The Deutsche Nationalbibliothek lists this publication in the Deutsche Nationalbibliografie; detailed bibliographic data are available in the Internet at http://dnb.d-nb.de.

ISBN 978-3-944331-36-2

Verlag der Wissenschaften GmbH
Bergstr. 70 | D-01069 Dresden
Tel.: 0351/47 96 97 20 | Fax: 0351/47 96 08 19
http://www.tudpress.de

Gesetzt vom Autor. | Typeset by the author.
Printed in Germany.

Experimental Investigations of Platform Film-Cooling in Highly-Loaded Turbine Vanes

A thesis accepted by the Faculty of Engineering of the Technische Universität Dresden in partial fulfilment of the requirements for the degree of Doctor of Engineering Sciences (Dr.-Ing.)

by

Dipl.-Ing. Martin Kunze

born in Magdeburg

1st Reviewer:	Prof. Dr.-Ing. Konrad Vogeler
2nd Reviewer:	Prof. Dr.-Ing. habil. Bernhard Weigand
Date of defence:	30.09.2013

Institute of Fluid Mechanics
Technische Universität Dresden
2013

Preface

The present thesis work has been conducted at the Chair of Turbomachinery and Jet Propulsion (TUS) of the Technische Universität Dresden during my employment as research engineer. Furthermore, the research project, that delivered most of the experimental results, has been financially supported by Siemens Power Generation and the reported work would not have been possible without their continuous support over the years. The permission to publish this work is gratefully acknowledged.

First and foremost, I would like to express my sincere gratitude to my supervisor Prof. Dr.-Ing. Konrad Vogeler. He placed the trust in me to grow into the challenges that would occur along the way and followed my work with much interest. Prof. Vogeler was very helpful during the investigations giving efficient guidance from his immense technical background in gas turbines whenever needed and receives the authors utmost respect and admiration. The same is true in regard to Prof. Dr.-Ing. habil. Bernhard Weigand from the Institute of Aerospace Thermodynamics (ITLR) at the Universität Stuttgart who was always willing to share his tremendous expertise in the domain of film cooling and heat transfer. This thesis profited a lot from his experience, knowledge and unique guidance. In appreciation of his encouragement and helping the thesis through the approval process at Siemens, I would like to thank Prof. M. E. Crawford for his efforts who besides made it possible that several research papers could be published in the ASME Journal of Turbomachinery.

I am very grateful to my colleagues at TUS for the cheerful environment and their assistance throughout the preparation of this thesis, in particular to mention the experimental work group. I will miss our daily lunch times and the subsequent coffee breaks where we shared lots of fun and laughter. Especially, I would like to give particular thanks to my colleagues Dipl.-Ing. Andreas Krug and Dipl.-Ing. Peter Busse with whom I had the pleasure to share an office for many years. I appreciate the exiting times during our „friday afternoon sessions“ and the growing friendship that arose from your company. The fruitful discussions about everything and anything had certainly brightened our labor days. I wish you all the best for your own thesis and the ongoing work. Furthermore, special thanks go to Eberhard Wolf who had thrown in his enormous knowledge about pressure and temperature measurement techniques to set up the data acquisition systems for both test rigs. Particularly in regard to his tireless ambition and motivation to discuss the secondary loss results, I would like to thank Dr. sc. techn. Helmut Sauer for his guidance throughout the measurement campaign. Despite his retirement, he was willing to share his extensive knowledge on secondary flow. Furthermore, I would thank my students Tilman auf dem Kampe who took charge of the numerical model and Stefan Preibisch who participated in the flat plate experiments. Equally so, I like to give sincere thanks to all staff of the mechanical and electrical workshop department, represented by Christiana Sperling, Joachim Koch, Egbert Kretzschmar and Thomas Hoffmann. None of the test rig's constructive realization would have been possible without their engineering and technological experience.

Finally, I wish to express my greatest and deepest thanks to my family. The foundation for the present work and its success has been laid by my parents who raised me to the person I am today and gave permanent support and confidence throughout my education. In particular, I would like to thank my wife Corinna and my little son Julian for their patience, understanding and enduring support during long working hours and weekends absent from home to prepare this thesis.

Martin Kunze
Dresden in November 2013

Contents

Nomenclature

Latin Characters

Symbol	Unit	Description
A	[m^2]	cross-sectional area
A_R	[$-$]	hole area ratio, $A_R = A_{fan}/A_{met}$
c	[mol]	concentration
c_p	[J/(kgK)]	specific heat capacity
C_D	[$-$]	discharge coefficient
C_p	[$-$]	static pressure coefficient
Cov	[$-$]	coolant wall coverage
D	[m]	diameter of film-cooling hole
D_R	[$-$]	coolant-to-mainflow density ratio
e	[m]	rib height
e	[J]	internal energy
E	[J]	total energy
E_A	[J/mol]	Arrhenius activation energy
f	[m]	focal length
F	[N]	force
h	[J/(kgK)]	specific enthalpy
h	[$W/(m^2K)$]	heat transfer coefficient
$\hbar$	[$-$]	Planck's constant
H	[$-$]	shape factor of boundary layer
H	[m]	height
i	[°]	incidence
I	[$-$]	coolant-to-main flow momentum flux ratio
I	[W/m^2]	luminescent intensity
k	[m^2/s^2]	turbulent kinetic energy
k	[$mol/(m^3s)$]	chemical reaction rate
K	[$-$]	constants for the calibration of 5-hole probes
L	[m]	length
$\dot{m}$	[$-$]	mass flow rate
M	[$-$]	blowing ratio
Ma	[$-$]	Mach number

N	[−]	number of holes
p	[Pa]	pressure
p	[−]	order of convergence
P	[m]	pitch
Pr	[−]	Prandtl number
$\dot{q}$	[W/m^2]	specific heat flux
$\dot{Q}$	[W]	heat flux
r	[−]	recovery factor
r	[−]	grid refinement ratio
R	[m]	radius
R	[J/(kgK)]	specific gas constant
R_u	[J/(kgK)]	universal gas constant
Re	[−]	Reynolds number
s	[m]	stream-wise distance
S	[−]	sensitivity coefficient
S_{eff}	[m]	effective film hole slot width, $S_{eff} = \Pi/4D^2/P_h$
S_f	[−]	safety factor
t	[s]	time
t	[m]	thickness
T	[K]	temperature
T_R	[−]	coolant-to-main flow temperature ratio
Tu	[−]	turbulence intensity
u	[m/s]	velocity
u^+	[−]	non-dimensional velocity
$\overline{u}$	[m/s]	time-averaged velocity
$u^{'}$	[m/s]	fluctuating velocity
U	[V]	voltage
v	[−]	profile velocity ratio, $v = u_{m,2}/u_{m,1}$
V	[m^3]	volume
$\dot{V}$	[m^3/s]	volumetric flow rate
V_R	[−]	coolant velocity ratio, $V_R = u_c/u_m$
w	[m]	width
x	[m]	axial coordinate
y	[m]	pitchwise coordinate
z	[m]	spanwise coordinate
x_p, y_p	[m]	profile coordinates
x_S, y_S	[m]	streamwise-oriented cartesian coordinates
z^+	[−]	non-dimensional wall distance
Z	[−]	compressibility factor

Greek Characters

α	[°]	streamwise inclination angle of the film hole
α_M	[m^2/mol]	molar extinction coefficient
β	[°]	compound angle of the film hole
β	[°]	flow angle in x-y plane
γ	[°]	diffuser opening angle
Γ	[1/s]	vorticity
δ	[°]	flow angle in y-z plane
δ^*	[m]	displacement thickness
Δ	[−]	difference
ϵ	[m^2/s^3]	turbulent dissipation
ζ	[−]	total pressure loss coefficient
η	[−]	film-cooling effectiveness
Θ	[m]	momentum thickness
Θ	[−]	non-dimensional temperature
κ	[−]	isentropic exponent
λ	[W/(mK)]	thermal conductivity
λ_c	[m]	filter cut-on wave length
μ	[kg/(ms)]	dynamic viscosity
ν	[m^2/s]	kinematic viscosity
ξ	[−]	elemental error parameter
Π	[−]	pressure ratio
ρ	[kg/m^3]	density
σ	[−]	standard deviation
τ	[Pa]	shear stress
Φ	[W]	light power
Φ	[−]	quantum yield of luminescence
ψ	[kg/(ms)]	dilatational viscosity
ω	[1/s]	turbulent frequency

Subscripts and Superscripts

0	without film-cooling injection
$1, 2...4$	measurement locations at cascade
a	absolute, absorption, area-averaged
amb	ambient
av	averaged across film row
aw	adiabatic wall
ax	axial

c	cooling
cal	calibrated
cas	cascade
cl	centerline
$cond$	conduction
$conv$	convective
$corr$	correction
cyl	cylindrical
dis	discharge
D	based on hole diameter
e	emission, exit
ew	endwall
ex	extinction
exp	experiment
f	film
fan	fan-shaped
FE	finite-element
g	gas
h	hole
i, j	counting indices
i	inlet
ibl	inlet boundary layer
is	isentropic
l	local, left
lat	lateral, laterally-averaged
m	main flow
m	mass flow-averaged
mix	mixed gas
ms	midspan
M	measured value
n	normalized
p	pressure
pit	pitot
pl	plenum conditions
pr	Prandtl probe
P	profile
$offs$	offset
OP	operating point of vane
r	rib, right

rec	recovery
ref	reference
s	standard condition
S	secondary flow
t	total, turbulent
th	theoretical
w	wall

Abbreviations

CCD	Charged-Coupled Device
CFD	Computational Fluid Dynamics
CNC	Computer Numerical Control
CO_2	Carbon dioxide
CTA	Constant Temperature Anemometry
CV	Corner Vortex
CVP	Counter-Rotating Vortex Pair
DMM	Digital Multimeter
EDM	Electrical-Discharge Machining
FE	Finite-Element Method
FS	Full Scale
GCI	Grid Convergence Index
HSV	Horse-Shoe Vortex
IEEE	Institute of Electrical and Electronics Engineers
LE	Leading Edge
LED	Light-Emitting Diode
LFE	Laminar Flow Element
LSRCW	Low Speed Research Cascade Wind Tunnel
MCR	Mass Flow Controller
MFR	Mass Fraction Ratio
ML	Measurement Location
MP	Measurement Plane
PS	Pressure Side
PSP	Pressure-Sensitive Paint
PV	Passage Vortex
PVC	Polyvinyl chloride
RANS	Reynolds-Averaged Navier Stokes
SDIU	Scanivalve Digital Interface Unit
SS	Suction Side

SLPM	**S**tandard **L**iter **P**er **M**inute
SMD	**S**urface-**M**ounted **D**evice
SNR	**S**ignal-to-**N**oise **R**atio
SV	**S**olenoid **V**alve
SVO	**S**treamwise **Vo**rticity
TBC	**T**hermal **B**arrier **C**oating
TEC	**T**hermo-**E**lectric **C**ooler
TC	**T**hermo**c**ouple
TCV	**T**ip **C**learance **V**ortex
TE	**T**railing **E**dge
TSP	**T**emperature-**S**ensitive **P**aint
TV	**T**railing Edge **V**ortex

Abstract

Platforms of high pressure turbine vanes are subject to extreme aerodynamic and thermal loads. As a result, these surfaces are extensively cooled using the film-cooling technique. Here the coolant is injected through small holes in the platform walls into the near-wall flow field which is highly three-dimensional due to secondary vortex structures. These vortical structures are driven by the radial and circumferential pressure gradient which change with the gas turbine's operating point. While the inlet flow conditions of the vanes are altered during start-up or in part load, the increased loading directly affects the strength of the secondary flow and the coolant-hot gas interaction. With the difficulties in predicting the film-cooling behavior under the influence of the secondary flow with the use of existing correlations. Thus, experimental investigations are required to gain physical understanding of the complex coolant-hot gas interaction near the platform walls. Furthermore, it provides an experimental data base with variation of airfoil loading to develop correlations that can be integrated into the design system of gas turbines.

The present work focuses on the experimental investigation of the platform cooling configuration of a highly-loaded gas turbine vane. Different loading conditions are applied to a straight vane cascade by changing the airfoil's incidence. The cooling performance is discussed in terms of the spatial and streamwise development of the film effectiveness. For the determination of the surface temperature fields, a non-intrusive, optical measurement technique based on temperature-sensitive paints (TSP) is used. This implies real temperature gradients between the main flow and the coolant which induce three-dimensional conduction effects within the test plate. Hereby the measured wall temperature is altered significantly. Therefore, a thermal analysis based on a finite element method was used to simulate the conductive heat fluxes and to correct the temperature measurements.

The scope of the experimental investigations is subdivided into two parts. The first part of the present work involves the validation of the TSP-technique on a basic flat plate film-cooling configuration. Results of the film effectiveness are obtained for a cylindrical and a fan-shaped hole geometry with the variation of blowing rate and density ratio. Centerline and laterally-averaged film effectiveness for both film hole geometries are compared to previous studies. The centerline film effectiveness show a reasonably good agreement with the previous studies as long as the film jets stay attached to the wall. The differences slightly increase with increased coolant momentum. They are mainly attributed to different fluid and geometry boundary conditions. The laterally-averaged film effectiveness is always higher compared to the literature data. The differences may be attributed to lateral conduction effects induced by the hot films.

During the second and main part, experimental investigations were carried out on the airfoil cascade with endwall cooling. Prior to coolant injection, the aerodynamic performance of the vane cascade was investigated. The total pressure loss results reveal a doubling of the endwall loss coefficient by increasing the incidence from nearly zero to +40°. Furthermore, the secondary flow vectors and the streamwise vorticity show enhancement with elevated incidence. Strongly enhanced lateral pressure gradients being the driving force for the secondary flow are visible in the endwall and airfoil pressure distribution. Based on these findings, three test cases are selected for the endwall film-cooling. The reference case is set at nearly zero incidence to simulate the operating point of the real-engine vane. The other cases are selected with +15 and almost +40° incidence.

Endwall film-cooling is investigated for three single hole rows located at different positions as well as for a group of five rows located near the pressure side. The results reveal that the rows in the front part of the passage experience strong influence by the secondary flow. The effect on the rows in the rear part is less severe. Directly downstream of the hole exits, the film-cooling jets are deflected away from their initial trajectories which is dictated by the hole orientation. The coolant is transported in streamwise direction along with the near-wall flow pattern toward the suction side of the adjacent vane. Here, it impinges onto the profile surface and is entrained into the vortical movement of the secondary flow. With increased loading, more pronounced diversion of the coolant layer is found. This reduces the endwall coverage along with a shift of the impingement region in upstream direction. Also, the streamwise mixing of the coolant with the mainflow is significantly enhanced. This results into up to 50 % faster decay of the film effectiveness in streamwise direction.

Kurzzusammenfassung

Die Plattformen einer ersten Leitreihe der Hochdruckturbine sind sowohl hohen aerodynamischen als auch thermischen Belastungen ausgesetzt. Dazu werden sie mit Hilfe der Filmkühlungstechnologie umfangreich gekühlt. Hierbei wird Kühlluft durch kleine Bohrungen in den Wänden in die heißen Verbrennungsgase eingeblasen. Aufgrund der sich in der Schaufelpassage entwickelnden Sekundärwirbel ist die Plattformwandgrenzschicht hochgradig dreidimensional. Diese Wirbelsysteme werden von dem radialen Druckgradienten wie auch dem Druckgradient in Umfangsrichtung angetrieben. Änderungen des Betriebspunktes der Gasturbine, wie z.B. beim An- und Abfahren oder im Teillastbetrieb, haben somit einen Einfluss auf das Strömungsfeld in den einzelnen Stufen als auch auf die Belastung der Schaufeln. Dies wiederum beeinflusst die Stärke der Sekundärströmungswirbel und damit die Interaktion zwischen Kühlluft und Heißgas. Aufgrund der Schwierigkeit, den Einfluss der Sekundärströmung auf die Filmkühlung mit Hilfe von bestehenden Korrelationen korrekt vorrauszusagen, sind experimentelle Untersuchungen hierzu unerlässlich. Diese erweitern den Kenntnisstand über das komplexe Zusammenwirken von eingeblasener Kühlluft und heißgasseitiger, wandnaher Querströmung. Zusätzlich liefern die Untersuchungen eine Datenbasis in Abhängigkeit von der Schaufelbelastung, auf deren Basis Korrelationen für das Auslegungssystem von Gasturbinen entwickelt werden können.

Die hier vorgestellte Arbeit konzentriert sich auf experimentelle Untersuchungen zur Filmkühlung an den Seitenwänden einer hochbelasteten Gasturbinenleitschaufel. Die Veränderung der Schaufelbelastung im geraden Schaufelgitter erfolgt durch die Variation der Inzidenz. Das Kühlluftverhalten auf der Seitenwand wird anhand der örtlichen Verteilung der Filmkühleffektivität sowie deren Entwicklung entlang der Strömungsrichtung diskutiert. Zur Erfassung der Oberflächentemperaturverteilung kommt ein berührungsloses, optisches Messsystem zur Anwendung, welches auf temperatursensitiven Farben (TSP) basiert. Hierzu ist die Bereitstellung eines Temperaturgradienten zwischen Kühlluft und Hauptströmung notwendig, was zu Wärmeleitungseffekte in der Seitenwandplatte führt. Trotz der Verwendung von Werkstoffen mit geringer aber endlicher Wärmeleitfähigkeit werden die gemessenen Temperaturverteilungen verfälscht. Zur Bestimmung der konduktiven Wandwärmeströme, werden thermische Analysen auf Basis der Finite-Elemente-Methode durchgeführt. Die daraus erhaltenen numerischen Ergebnisse werden zur Korrektur der Rohmessdaten in den Auswerteprozess integriert.

Der Umfang dieser Arbeit gliedert sich im Wesentlichen in zwei Teile. Im ersten Teil wird die TSP-Messtechnik an einem Filmkühlaufbau an der ebenen Platte erprobt und validiert. Hierzu werden Filmkühleffektivitätsverteilungen an einer zylindrischen und an einer Diffuserlochgeometrie mit Variation der Ausblasrate und des Dichteverhältnisses gemessen. Im Fall der zylindrischen Löcher zeigt der Vergleich der Filmeffektivitäten entlang der Strahlmittellinie eine sehr gute Übereinstimmung mit den Ergebnissen aus der Literatur solange die Filmstrahlen noch an der Wand anliegen. Bei Erhöhung des Austrittsimpuls zeigen sich leicht abweichende Ergebnisse, die aber durch die unterschiedlichen experimentellen Randbedingungen bedingt sind. Die Diffuserlöcher zeigen indes für die Filmeffektivitätswerte entlang der Mittellinie eine gute qualitative und quantitative Übereinstimmung mit den Literaturdaten. Die teilungsgemittelten Größen sind dabei immer etwas höher als die Ergebnisse der vergleichenden Studien. Dies ist auf laterale Wärmeströme zurückzuführen, die durch den heißen Strahlkern initiert werden.

Innerhalb des zweiten Arbeitspaketes wird das Turbinenschaufelgitter mit Plattformkühlung für unterschiedliche Betriebspunkte untersucht. Vorab erfolgen detaillierte Messungen zum Strömungsfeld vor und hinter dem Gitter. Die Ergebnisse für den Totaldruckverlust zeigen mit Variation der Schaufelinzidenz von null bis auf 40 Grad eine mehr als Verdopplung des Sekundärströmungsverlustes. Dies deutet auf eine beträchtliche Intensivierung der Sekundärströmung hin. Auch die Sekundärströmungsvektoren und die Wirbelstärke in Strömungsrichtung unterstützen diese Schlussfolgerung. Weiterhin zeigen die Seitenwand- als auch die Schaufeldruckverteilungen stärkere laterale Druckgradienten mit höherer Inzidenz. Auf Basis dieser Erkenntnisse wurden drei verschiedene Belastungsfälle für die Filmkühlversuche festgelegt. Der Referenzfall wird bei einer sehr geringen Inzidenz durchgeführt. Dies entspricht dem Betriebspunkt der ersten Leitreihe in der Maschine. Die beiden anderen Testfälle bei höherer Profilbelastung werden bei $+15°$ beziehungsweise $+40°$ Inzidenz durchgeführt.

Die Filmkühlung an der Gitterseitenwand wird anhand drei einzelner Bohrungsreihen mit unterschiedlicher Lage in der Gitterpassage sowie für eine Gruppe aus fünf Bohrungsreihen nahe der Schaufeldruckseite diskutiert. Die Ergebnisse zeigen einen erheblichen Einfluss der wandnahen Querströmung auf die Ausbreitung und Verteilung der Kühlluft. Dies gilt vor allem für die Lochreihen die im vorderen Teil der Schaufelpassage wie beispielsweise am Gittereintritt oder im direkten Einflussbereich der Schaufelvorderkante positioniert sind. Hierbei erfahren die Filmstrahltrajektorien direkt nach dem Lochaustritt eine starke Ablenkung von ihrer vorgegebenen Einblasrichtung und werden infolge des wirkenden Querströmungsgradienten in Richtung Saugseite der benachbarten Schaufel transportiert. Dort treffen die Kühlluftstrahlen auf den Schaufelfuss. Sie werden dann durch die Querströmung von der Seitenwand abgehoben und in die entstehende Wirbelbewegung integriert. Mit steigender Belastung und somit stärkerem Querdruckgradienten im Gitter wird die Sekundärströmung intensiviert und damit die Krümmung der wandnahen Stromlinien verstärkt. Folglich steigt die Ablenkung der Filmstrahlen und der Auftreffbereich auf der Schaufeloberfläche verlagert sich stromauf in Richtung Vorderkante. Weiterhin kommt es aufgrund der stärkeren Vermischung der Kühlluft mit der Hauptströmung zu einer stärkeren Abnahme der Filmkühleffektivität in Strömungsrichtung. Der Strömungsweg, den die Filme auf der Seitenwand zurückgelegen, verringert sich dabei um bis zu 50%. Damit wird auch die gekühlte Seitenwandfläche mit steigender Schaufelbelastung sichtbar kleiner.

1 Introduction

Nowadays, turbomachinery applications are widely used in our modern society and play a major role in our daily life. They are used as industrial gas turbines for power generation, as jet engines for aircraft propulsion, turbo chargers in the automotive industry or as industrial compressors. Although renewable energy sources gain even more importance in the field of power generation, fossil fuels still have to cover largely the demand for energy of a fast growing world population. During the last decades, growing requirements on thermal efficiency and power output restricted by lower noxious emission specifications and reduced noise limits have driven substantially the research and development of key technologies in turbomachines. This has led to remarkable advancements over the years in the field of aerodynamics, mechanical integrity, cooling design, combustion kinetics as well as material science.

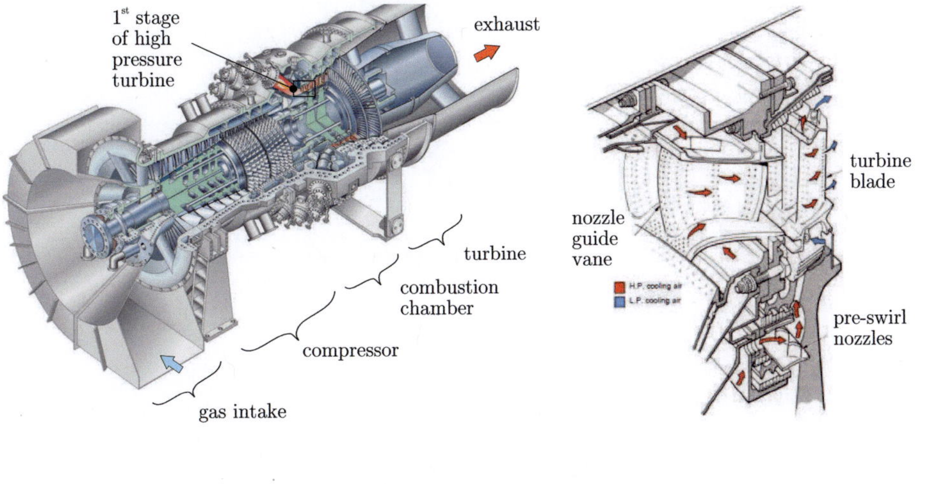

Figure 1.1: Sectional view of a modern heavy duty gas turbine (source: www.siemens.com) and schematic of a first stage of a high pressure turbine from a jet engine with coolant distribution from Rolls-Royce [107]

A gas turbine is composed of three main components denoted as compressor, combustion chamber and turbine (see also Fig. 1.1). The working principle of this type of machines is the thermodynamic Brayton cycle in which the thermal efficiency and specific work is influenced by the compressor pressure ratio as well as the turbine inlet temperature together with the efficiencies of the components. At the engine inlet, air at large mass fraction is drawn from atmosphere to enter the compressor. It consists of a low, medium and high pressure compressor part when engines with a typical three-shaft configuration are considered. The fluid is compressed to high pressure levels by a series of stator and rotor blade rows. It also results in a significant increase in temperature. The high-pressure gas is then passed into the combustion chamber to be mixed with the injected fuel and be burned. Subsequently, the high-temperature and high-pressure fluid enters the turbine to expand to atmospheric pressure over several stages of stator and rotor blades. The work gained in the turbine is mainly used to drive the corresponding compressor part which is mounted on the same shaft. The remaining energy is either used for aircraft propulsion or for powering a generator mounted on the same shaft to produce electrical energy.

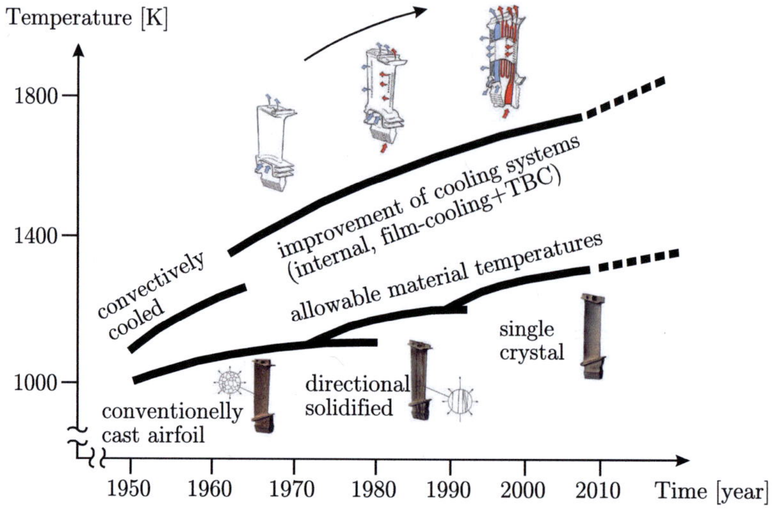

Figure 1.2: Development of cooling systems and materials for turbine blades related to the turbine inlet and material temperature

The thermal efficiency is furthermore affected by the required amount of cooling and leakage air used in the turbine. Due to the continuous increase of the inlet temperature of the turbine, modern high pressure turbine components operate well above the limiting material temperatures. In particular, the first guide vane of the high pressure turbine is subject to withstand increasing thermal and aerodynamic loads as the hot gas directly exits the combustion chamber. Therefore a large amount of cooling is provided to this specific part of the turbine. Here the turbine platforms and the blading are critical regions related to excessive thermal failure. A combination of different method like internal convective cooling and film-cooling is used to cool the turbine components. To enhance the heat convection inside the cooled component, internal cooling structures are designed as multi-pass channels with implemented ribs and pin-fin arrays. The development of new high-temperature resistant alloys and the continuous improvement of the casting process of the turbine airfoil, e.g. from a conventionally cast blade to a single-crystal piece, has lead to materials with significantly enhanced thermal strength that can be used in the hazardous turbine environment. However the increase of the maximum material temperature is smaller than the consistent trend of increasing the turbine inlet temperature. Its evolution over the last decades is sketched in Fig. 1.2.

In order to reduce the metal temperatures to levels that prevent airfoils from suffering thermal failure and guarantee the designated life time of the engine, complex cooling methods must be applied to the turbine components. The continuous rise in the turbine inlet temperature during the last four decades is mainly facilitated by the advancement of the blade cooling systems. Modern turbine parts are designed using nickel-based super alloys while the surfaces are covered with low conductive thermal barrier coatings. Further the internal airfoil structure incorporates complex multi-pass cooling systems with numerous cooling holes. Therefore, a large amount of cooling and bleed air must be provided at the compressor exit which is not progressed to the combustion process. A modern gas turbine with turbine inlet temperatures of typically 1770 K consumes about 18 to 20 % of the mass flow at the compressor inlet as stated by Vogeler [128]. Here, the first guide vane receives the major part of about 6 % of the inlet mass flow which equals approximately 25 % of the entire coolant consumption of the turbine.

1.1 Cooling Systems in Gas Turbine Airfoils

Modern turbine airfoils with multi-pass cooling channels are highly complex and cost-intensive cast parts. They employ a combination of internal convective cooling structures and film-cooling holes (see also Fig. 1.3). For the use of film-cooling, the turbine parts are perforated to provide sufficient coolant injection onto the wall surfaces that are exposed to the hot combustion gases. Cold air is bled from the last compressor stages and ducted through the secondary air system and the shaft section into the supplying cavities underneath the airfoil root. From there, a part of the coolant is injected into the wall boundary layers of the hot gas path through small, discrete holes incorporated in the hub and casing platform walls. The rest of the cooling air is routed into the multi-pass cooling system that is cast into the airfoil. Herein it is distributed into different radial channels with varying shape and aspect ratio being connected by 180° bends.

In a similar way, small holes of different length and shape are also integrated into the airfoil walls. The released cooling fluid generate a protective film around the airfoil surfaces displacing the hot combustion gases from the wall. Thus, it reduces the driving temperature difference for the heat transfer between the hot gas and the blade wall. As a side-effect, heat is drawn from the metal parts by convective heat transfer inside the film-cooling hole while the coolant

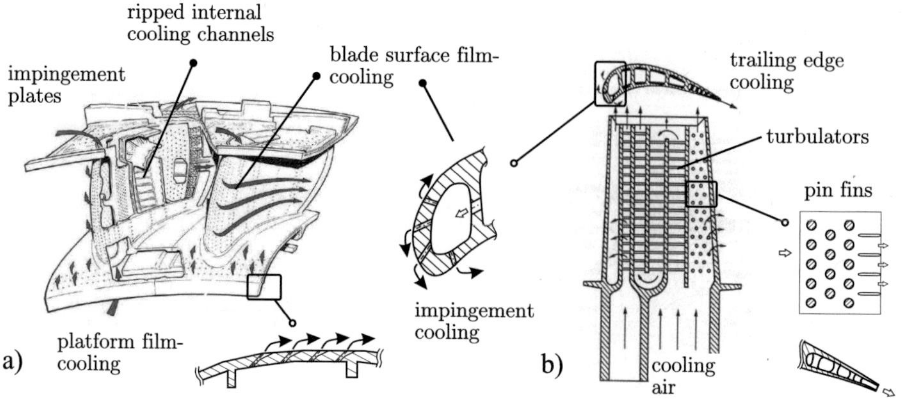

Figure 1.3: Cooling methods in high pressure turbine components: a) nozzle guide vane of a jet engine adapted from Rolls-Royce [107] and b) rotor blade of first stage adapted from Han [50]

is convected through it. This leads to a significant reduction of the material temperature. With the application of cooling methods, the amount of fuel can be increased which leads to greater power output along with the drawback of higher thermal loads of the turbine components.

Cylindrical hole geometries were commonly used in the past in turbine cooling systems for all different kind of cooling aspects. During the last decades this type of hole geometry has been replaced by an advanced cooling-hole design using shaped hole exits. Here the hole contains a cylindrically formed inlet part and the exit portion is shaped as a diffuser over the last two or three hole diameters. Due to the increased hole cross-section in lateral and streamwise direction, the exit momentum of the film jet is sufficiently reduced. This leads to enhanced lateral wall coverage by the cooling films but also to improved stability of the films with variation of the operating conditions. It further results into less dependency of the jet development from the blowing parameter. This improves and stabilizes the cooling performance over a larger range of coolant mass flows. Due to these advantages, shaped holes are preferred as cooling hole geometry in modern cooling designs. They are used not only on the airfoil surfaces but also to a great extend on hub and casing platforms.

Further reduction of the wall heat flux can be achieved by application of insulating coatings (TBC). Ceramic material with low heat conductivity is applied to all wall surfaces that are exposed to the hot gases. Those coatings consist of four layers: the metal substrate, a metallic bond coat, the thermally-grown oxide and the ceramic top coat which is typically composed of zirconium-based ceramics. The extracted cooling amount is not proceeded to the combustion which involves the loss of power output and thermal efficiency. Therefore the primary goal of the cooling system's design process is to reduce the required amount of coolant by taking into account the component life time and the required process parameters. Further optimization of the coolant distribution in the turbine improves the efficiency of the cooling system.

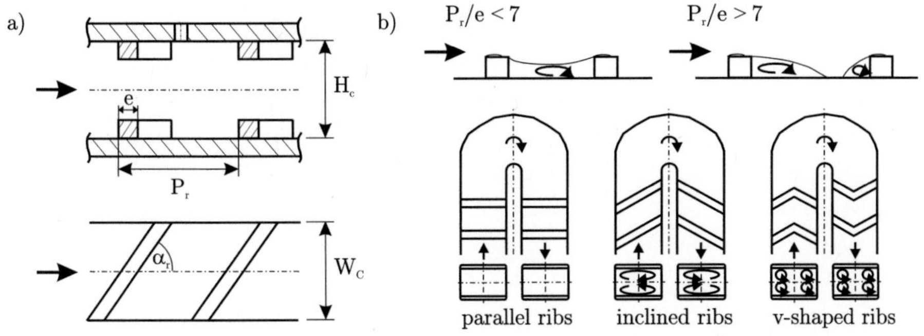

Figure 1.4: Geometrical parameter of ripped cooling channels (a); schematic of flow field induced by cooling ribs with variation of pitch-to-rib height ratio P_r/e and different rib configurations with corresponding secondary flow field in rectangular cooling channels (b)

For the convectively-cooled inner part of the airfoil, the design efforts are aimed at maximizing the internal heat transfer. Their primary goal is to extract as much heat as possible from the blade material. This reduces the metal temperature to meet the specified design limits which guarantees the component's life time. Turbulence generators inside the internal cooling system contribute to the increase of heat transfer (see Fig. 1.4 a). The extremely high heat transfer at the stagnation region around the leading edge significantly increases the blade's thermal load. Therefore, the triangularly-shaped cooling channels at the leading edge region are equipped with impingement pans for vanes. For rotating blades, the holes are implemented into the casting core of the airfoil. These arrays produce small impinging jets onto the inner surface of the leading edge wall which increase the local heat transfer. The jet flow field is associated with a stagnation point and is composed of three parts: the potential inner core of the jet, the shear layer with the surrounding flow due to viscous forces and the wall jet that is created when the fluid is deflected at the wall in opposite directions. Since typical impingement arrays consist of multiple jets, the ideal jet structure is altered by jet-to-jet interactions.

In the region of the trailing edge, the backward flow channels extend toward the trailing edge opening. These channels can be equipped with arrays of circular pins or other differently-shaped protuberances (commonly denoted as pin-fin cooling). The function of the pin fins is to produce high heat transfer rates by periodically disturbing the channel flow. The heat transfer can be enhanced further by applying multiple, staggered rows of pins. During the pin-fin design process, the major geometric parameters like pin height, spacing, stagger or channel-height-to-length ratio must be selected appropriately to avoid high friction losses.

With the implementation of rib turbulators onto internal cooling-channel walls, the hydrodynamic and thermodynamic boundary layer is periodically disturbed by the placed protrusions. This leads to periodic separation and reattachment zones upstream as well as downstream of the ribs and to increased friction losses inside the cooling channel. It depends on the pitch-to-rib height ratio whether the flow reattaches inside the rib segment or not. This produces the re-start of the thermal boundary layer which is associated with elevated heat transfer. One drawback of the heat transfer enhancement achieved by ribs is a significant rise of the pressure drop along the channel. It can be several times larger than in a smooth channel. Additionally the turbulence of the core flow in the cooling channel is increased which further enhances the heat transfer.

The heat transfer performance, the friction losses and the resulting aerodynamic flow field of a ribbed cooling channel depend on the channel aspect ratio, the rib shape, the number of ribs, the height, pitch ratio and angle orientation as well as the coolant Reynolds number. To further increase the heat transfer, the transverse ribs can be oriented with an angle of attack lower than 90° with regard to the coolant core flow. The highest heat transfer coefficients accompanied with the highest friction losses are produced with 60° inclined ribs. Such angled rib configurations induce different secondary flow structures that additionally alter the heat transfer situation. Various rib configurations as v- or w-shaped ribs are common (shown in Fig. 1.4b). They result in different formations of the secondary flows and hence changing influences on the heat transfer rates.

The implementation of cooling holes is done after the casting process by either laser drilling, water jet drilling or electrical-discharge machining (EDM). The selection of the appropriate method mainly depends on desired quality and size of the hole shape. Less attention is paid to the positioning of the holes with regard to the rib segment and the rib edges. Thus, it is obvious that holes can be placed near the stagnation point of the cooling flow or in the separation zones behind the ribs. This may result in flow field changes at the inlet and inside the film-cooling holes. A reduced distance between the rib and the hole inlet lead to decreased discharge coefficients of the film holes (Heneka [55]). The implementation of ribs in the cooling channel enhances the heat transfer coefficient on the hot gas path with film injection compared to a smooth channel (Kissel [69]). Changes in film effectiveness due to ribs are less significant.

2 Film-Cooling in Turbine Airfoils

The film-cooling process in modern gas turbine systems is not only applied to the airfoils which are subject to thermal failure. It is also used on the platforms to avoid hot-spots caused by coolant that is swept off the surface by the near-wall vortical flow structures. Furthermore, cooling is applied by leakage flows to the intersections of two vanes and to the tip gap section of the blades. The physical process is of three-dimensional, turbulent and unsteady character with respect to the thermal and flow field. The platform cooling holes are supplied from cavities underneath the endwalls. Here, the coolant pressure is mainly stable but the pressure ratio and hence the blowing parameter varies significantly across the passage.

The coolant is injected into the distorted wall boundary layer with its stream-wise trajectories being dominantly influenced by the near-wall flow field. Cooling of the airfoil is also done by a large number of holes placed over a large portion of its span. The cooling holes are fed by a internal cooling supply system which consists of differently shaped channels. The internal features increase the pressure loss along the channels which results in decreasing coolant pressures in spanwise direction. Also the flow field inside the channels is significantly altered by inserted turbulators which in turn influences the aerodynamics inside the cooling holes. Since the film-cooling method incorporates two fluids at different temperature levels and velocities which involves strong interaction and mixing, the process depends on a great number of parameters. Therefore, the following sections only intend to give a short insight into the physical origins of film-cooling and its influencing parameters.

2.1 Main Gas Path

2.1.1 Secondary Flows in Turbine Bladings

The endwall of a turbine blading is a region complicatedly to be cooled for several reasons. Firstly, the endwall area represents a large fraction of the entire cooled vane surface which results into large coolant consumption for these regions. This is especially true for modern airfoil designs with low aspect ratio and low solidity. Secondly, a highly three-dimensional and unsteady flow field is present near the endwall induced by pressure and temperature gradients. It is dominated by several vortex structures wherein the individual phenomena are summarized as secondary flow. Injection of coolant with low to moderate momentum into this highly distorted wall boundary layer will lead to complicated interaction and mixing of the films with the cross flow. The mixing process depends on the characteristics and strength of the secondary flow. Furthermore, changes of the turbine's operating point and variation of the incoming flow from the combustion chamber has tremendous effect on the development of the secondary flow, hence on the cooling performance.

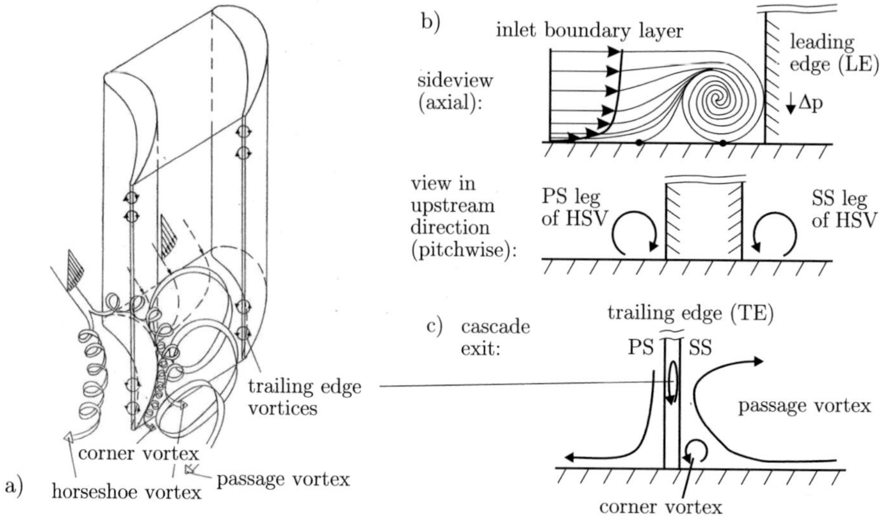

Figure 2.1: Secondary flow structures in a turbine cascade passage adapted from Sauer et al. [110] a), leading edge region: roll-up of inlet boundary layer at leading edge to form the horse-shoe vortex (HSV) b) and cascade exit: development of corner and trailing edge vortex c)

The term secondary flow refers to a superposition of different phenomena which are not part of the two-dimensional main flow at midspan. Several vortical structures develop in the near-wall region and interact with each other. The most dominant flow structure is the passage vortex (see Fig. 2.1) followed by the horseshoe vortex which is developed around the leading edge, corner vortex and trailing edge vortices. The passage vortex is driven by the radial pressure gradient and is formed by the crossflow of fluid inside the low-energy endwall boundary layer from the pressure side (PS) of a blade to the suction side (SS) of the adjacent blade. The near-wall crossflow interacts with the profile boundary layer at the rear-part section of the airfoil in the corner region between the endwall and the airfoil suction side. It heavily disturbs the newly developing profile boundary layer which results into increased total pressure losses and eventually local flow separation in those airfoil regions.

The horse-shoe vortex is developed around the leading edge due to the spanwise pressure gradient. In the stagnation region, the pressure rises significantly in streamwise direction. Therefore the wall-near layers of the inlet boundary layer are moving toward mainflow regions with lower pressure and begin to roll-up into spiral vortices. This forms two different branches with opposite sense of rotation which are transported into the adjacent passages. It is denoted as suction side and pressure side branch of the horse-shoe vortex. The weaker SS-branch of the horse-shoe vortex develops close to the corner between the airfoil suction side and the endwall as it is swept downstream. In contrast, the stronger pressure side leg is driven across the passage by the pitchwise pressure gradient and becomes an integral part of the passage vortex.

In the rear part of the airfoil, the highly distorted near-wall cross flow impacts on the suction side with a nearly wall-normal velocity component and is diverted toward midspan. Due to the shear layer that is created between the vortical flow and the fluid inside the suction side corner, the flow rolls up to form a corner vortex with an opposite sense of rotation. Due to its spatial extent that is far smaller than the passage vortex and its weak vorticity, it can hardly be visualized in aerodynamic experiments. However, its influence on the flow turning and the outlet flow angle is detectable.

As the flow convects through the passage and exits the airfoil cascade, the passage vortices from the two adjacent channels have opposite rotational movement. At the suction side, the vortex shows an upward movement towards midspan while at the pressure side, the flow is transported downward to the endwall. Thereby, this counter-rotating motion initiates pairwise discrete vortices inside the shear layer between the two flows, also denoted as trailing edge vortices. Their strength is mainly influenced by the strength of the passage vortex and therefore dependent on the airfoil geometry in terms of turning and velocity ratio.

In the rotating components of a gas turbine, the construction-required gap between the airfoil and the casing induces an additional secondary flow phenomenon. The flow field in the tip region of a blade is strongly influenced by the tip clearance vortex (TCV). The pressure difference between the two blade surfaces drives fluid to move through the clearance from PS to SS. Leaving the gap and entering the next passage, the fluid's interaction with the main flow inside the shear layers forms the roll-up into a spiral vortex. It is then transported across the passage. At elevated airfoil loading, the vortex may impact onto the adjacent pressure side depending on its driving pressure gradient. Further details on the periodical unsteady flow field near the blade tip and the interaction of the TCV and the mainflow as well as the incoming wakes of the upstream stator can be found in Mailach [91].

The classical secondary flow model was first described by Hawthorne [53]. It presents the vortex system as a composition of components of vorticity in flow direction. This is a result of the inlet flow containing vorticity passing through a airfoil cascade. This picture of the endwall flow situation has been changed during the past years to incorporate more details of the flow structures that were mainly observed by experimental investigations. Sieverding [119] gives a detailed overview on the secondary flow behavior in turbine blade passages covering the years up to 1985. In a recent review, Langston [76] summarizes the existing secondary flow models. It further covers the past research efforts done in the secondary flow field as well as in secondary

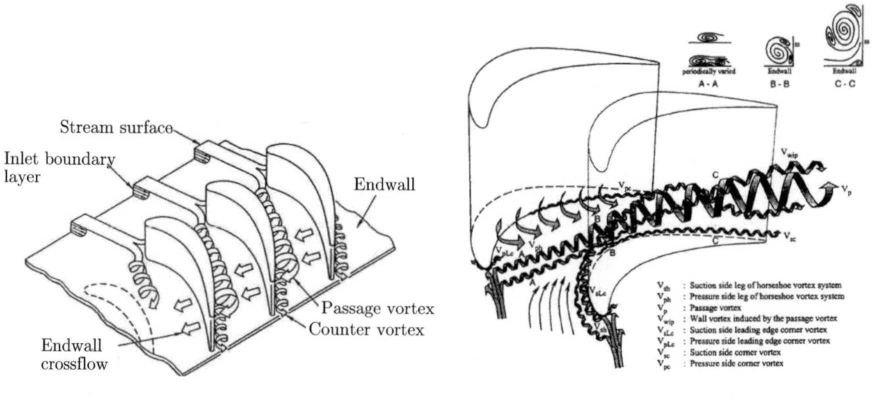

Figure 2.2: Secondary flow model by Langston [75] (left) and Wang et al. [132] (right)

loss reduction and prediction.

The model postulated by Langston [75] in 1980 (displayed in Fig. 2.2) clearly shows the separation of the inlet boundary layer at a saddle point to form the horse-shoe vortex. The PS-leg of the horse-shoe vortex merges with the passage vortex due to its identical sense of rotation. The suction side leg with its opposite sense of rotation continues inside the suction side endwall corner and is transported through the passage. It also introduces the development of a smaller vortex within the corner of the SS and the endwall. This corner vortex is driven by the shear layers between the upward movement of the passage vortex near the airfoil suction side and the low-moving fluid inside this endwall corner. It has opposite sense of rotation and the vortical motion is several times weaker than that of the passage vortex. Flow visualization studies carried out by Sieverding and Van den Bosch [120] using colored smoke wire techniques delivered a more detailed picture of the synchronous development of the horse-shoe and passage vortices. The findings of Sharma and Butler [118] instead showed a slightly modified pattern of the secondary flow with the suction side leg being wrapped around the passage vortex. Based on mass transfer measurements in a high-pressure turbine blade cascade, Goldstein and Spores [44] found a region of highly elevated mass transfer right at the leading edge-endwall junction. They presented therefore a revised secondary flow vortex pattern showing an additional leading edge vortex with small spatial extent but significant strength. The suction-side leg of the horse-shoe vortex remains close to the endwall and then lifts off when it reaches the separation line on the suction side. In a later investigation done by Wang et al. [132], a periodically fluctuating horse-shoe vortex system was observed. The vortices are traced by multiple smoke wires placed inside and outside of the endwall boundary layer and a laser light sheet that illuminated various cascade cross-sections. The secondary flow model further proposed the existence of the leading edge vortex which was also stated earlier in Goldstein and Spores [44].

2.1.2 Endwall Film-Cooling

There are numerous publications on endwall film-cooling. The subject has been investigated in terms of film effectiveness, heat transfer and aerodynamic performance. But little information can be found in the experiments about loading effects on endwall cooling. Detailed endwall film-cooling measurements were done by Friedrichs et al. [37–39] who successfully used the ammonium-diazo surface coating to measure film effectiveness for coolant injection through five rows of cylindrical holes distributed across the passage. The experimental results, as shown

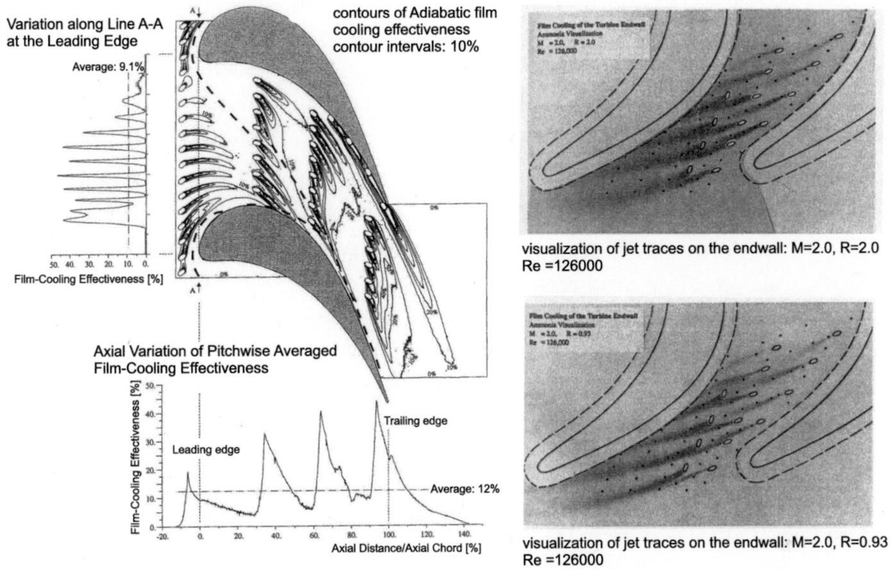

Figure 2.3: Film effectiveness distribution on a turbine endwall for injection through several rows of cooling holes measured by Friedrichs et al. [38] using the ammonium-diazo surface coating method (left), visualization of film jets on the endwall for two different density ratios D_R =0.93 and 2.0 at constant blowing rate M=2.0 studied by Jabbari et al. [58] (right)

in Fig. 2.3 left, indicate a strong interaction of the endwall coolant injection and the secondary flow in the passage. Their data showed that the coolant that is injected underneath the three-dimensional separation lines is inefficient due to strong mixing and decay of the jets as soon as they leave the holes. The jet trajectories of coolant that is injected directly into the near-wall cross flow were observed to be highly deflected toward the suction side and leaving pressure side regions essentially uncooled. Hence, the angle of injection only dictates the jet trajectories in the near hole vicinity. Furthermore, injection of coolant in the front of the leading edge region shows reduced lateral spreading and mixing. Therefore, it proceeds much longer downstream compared to the hole present in the passage. This provides improved cooling performance in streamwise direction which reaches at the SS-region downstream up to 30 % chord.

Zhang and Jaiswal [144] studied coolant injection onto the endwall surface by a staggered row of holes and a discrete slot in front of the cascade leading edge for various coolant-to-main flow ratios. Film effectiveness was measured using pressure sensitive paints (PSP) which is based on photoluminescence. This technique implies the analogy between heat and mass transfer wherein surface gas concentration at isothermal conditions is measured instead of directly obtain the surface temperatures. The authors found that the near wall flow field is dominated by the endwall secondary flow for low mass flow ratios. Thus, it results in strong deflection and quick decay of the coolant toward the suction side and ends up into low film effectiveness. With increased mass flow rates (above 2 % MFR), the secondary flow is suppressed and the coolant is able to move further to the pressure side. This is a result of the increased jet momentum.

Jabbari et al. [58] used an ammonium-diazo surface coating in the context of flow visualization. The authors investigated the behavior of jet traces injected through cylindrical holes distributed in a turbine blade endwall. The results show displaced jet centerlines away from its initial orientation. This effect is caused by the high deflection of the jets due to the strong cross flow present in this endwall region. The jet position is changed in lateral and radial direction with higher blowing rate and density ratio. At low density ratio of $D_R = 0.93$, stronger penetration of the low dense jets into the crossflow was observed. It leads to narrower traces on the endwall and low effectiveness.

Gaugler and Russell [40] compared visualized secondary flow pattern achieved by smoke flow and ink dot/oil drop methods with measured heat transfer distributions. The filmed endwall of a high pressure turbine vane in transonic flow was investigated by Kost and Nicklas [73, 99]. The authors investigated the aerodynamics, film effectiveness and heat transfer on the endwall with injection through discrete holes and an upstream slot configuration. Results indicate strong interaction of the injected coolant with the mainflow. The crossflow and subsequently the passage vortex is weakened both by coolant injection from the holes and the slot. Further, the horse-shoe vortex is intensified by the slot injection due to its position near the saddle point upstream of the blade leading edge. Slot injection provides high film effectiveness near the slot exit but strongly decreases toward the pressure side corner due to the interaction with the passage vortex. Strong increase in heat transfer was found due to the enhanced horse-shoe vortex. The weakened passage vortex causes a decreased heat transfer pattern.

Knost and Thole [65] studied two different film-cooling hole configurations combined with an upstream slot. The adiabatic film effectiveness was measured using an infrared-camera system. The results for the film-cooling injection show an uniform distribution of coolant through the passage. Interestingly, the results show an additional pattern which exhibits a large uncooled streak in the passage center. As stated by the authors, this hot streak exists in the region of the gutter and its leakage flow could alleviate this problem. The increase in coolant mass-flow rate results into higher film effectiveness levels and more effective cooling on the suction side. In combination with the slot injection, the results reveal a largely overcooled area at the cascade inlet. Predicted streamlines at 2% span were superimposed to visualize coolant trajectories. They were further compared to streamlines computed at midspan to reveal distinct differences. The authors stated that the near-wall flow structures should be considered when designing a new endwall cooling scheme.

Barigozzi et al. [6] compared cylindrical and shaped hole geometries on the endwall of a vane cascade using the thermochromic liquid crystal technique (TLC) to evaluate the film effectiveness. The shaped holes show a better aero-thermal performance than the cylindrical holes. This holds only true for low injection rates where the enhancement of the cooling performance is restricted to the front part of the vane passage. Larger improvements for the entire endwall can be observed with increased coolant mass flow rates. In the stagnation region around the leading edge, the cooling jets interact with the developing horse-shoe vortex. They are deflected away from their initial injection direction. Inside the cascade passage, the adiabatic film effectiveness distribution is mainly influenced by the near-wall cross flow. The optimum for the film-cooling coverage of the cylindrical holes is estimated at a MFR = 0.75 % (MFR ... mass fraction ratio) while the optimum for the shaped holes is found to be significantly greater at 1.5 %. The thermodynamic secondary-loss coefficient increases with larger MFR from 2.5 % to almost 6 %. The differences between cylindrical and shaped holes are small. In a later study, Barigozzi et al. [7] compared shaped cooling holes at two different free-stream turbulence intensities of 1 % and 9 %. The results of the thermodynamic secondary loss coefficient show that the area-averaged global losses are constant with increasing turbulence intensity. This holds also true when the coolant mass flow ratio was varied. Only small differences can be observed in the spatial film effectiveness distribution for both hole shapes. This indicates a negligible effect of the turbulence intensity on the cooling performance. Also the area-averaged film effectiveness showed very similar values for both turbulence intensities when comparing the results at constant MFR.

2.1.3 Endwall Heat Transfer

The endwall of an airfoil is characterized by distinct regions of potentially high heat transfer which are mainly influenced by the three-dimensional secondary-flow pattern near the endwall. Friedrichs [36] identified four main areas that are recognized by experimental results to show high heat transfer such as the leading edge, the suction surface shoulder, the pressure surface corner and the trailing edge wake. Furthermore, the author postulated three additional regions with high transfer: the suction side corner, downstream of the geometric throat and the lift-off line of the passage vortex. The extent of these regions seem to depend on Reynolds-number, the inlet boundary layer as well as the strength and position of the secondary flow. The resulting model as proposed by the author is shown in Fig. 2.4.

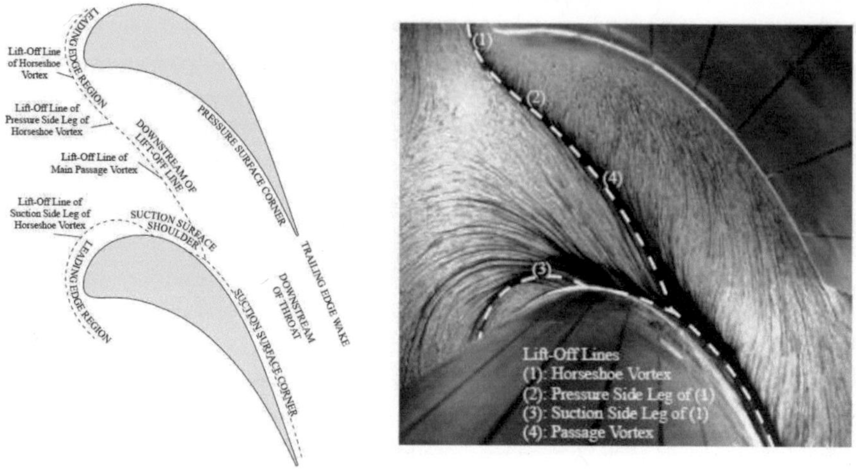

Figure 2.4: Subdivision of endwall regions with potentially high heat transfer (left) and visualization of the near-wall secondary flow pattern taken through oil-dye technique (right) from Friedrichs [36]

In most cases, when no boundary layer suction is applied, the incoming boundary layer to a turbine cascade has undergone a natural laminar-turbulent transition in the upstream wind tunnel components. Hence it is usually of turbulent character, has a large boundary layer thickness and exhibits low heat transfer. Therefore, the inlet region exhibits a wedge-like area of low heat transfer. It extends toward the passage mouth upstream of the three-dimensional separation lines.

Downstream of the separation lines, a new relatively thin boundary layer is developed. It is associated with high rates of heat transfer. The upstream region of the leading edge, bounded by the lift-off line of the horse-shoe vortex, is also characterized by high heat transfer. Its vortex formation is attributed to the roll-up of the incoming boundary layer with rotational movement toward the endwall (Blair [10], Gaugler and Russel [40]). The vortex transports hot main-flow material down towards the wall. In contrast, the suction-side surface-shoulder is the region where the near-wall cross flow, that has entrained the pressure side leg of the horse-shoe vortex, reaches the suction surface (Boyle and Russel [15], Takeishi et al. [124]). Here it interferes with the suction-side leg of the horse-shoe vortex and the newly developed airfoil boundary layer. This endwall region shows high rates of heat transfer which is probably due to interaction of the different vortex motions with the airfoil boundary layer. Near the trailing edge wake, the endwall region exhibits also high heat transfer coefficients (Graziani et al. [45], Goldstein and Spores [43]).

Heat transfer measurements and flow field visualizations done by Gaugler and Russel [40] indicate a downwash of fluid in the blade wake toward the endwall. Also Takeishi et al. [124] found elevated heat transfer downstream of the trailing edge caused by its strong wake which further enhances the heat transfer by high levels of turbulence being present in the wake. The pressure surface corner is also subject to elevated heat transfer (Haragasma and Burton [51]). Main flow material is swept down toward the endwall corner of the airfoil pressure surface to feed the near-wall cross flow that becomes the passage vortex. This vortical movement paired with elevated turbulence seems to attribute to the heat transfer enhancement in this region. Furthermore, the influence of the corner vortex on the heat transfer is noticeable (Takeishi et al. [124], Goldstein and Spores [43]). This vortex develops downstream of the throat region and is driven by the upward movement of the cross flow along the suction-side surface. It rotates in opposite direction with respect to the passage vortex and transports hot main fluid toward the wall. Additionally, high turbulence levels of the incoming flow can significantly augment the endwall heat transfer which was found by Kang et al. [70]. The authors state an increase in heat transfer of approximately 20 % at high inlet turbulence levels of 19.5 %.

Several authors have studied the heat transfer for stator endwalls. In 1974, Blair [10] conducted supposely the first investigation concerning film effectiveness and convective heat transfer in a turbine vane-passage for coolant injection through an upstream slot. The author concluded that the vane-passage heat transfer is strongly influenced by the vortex structure located in the corner between the endwall and the suction-side surfaces. Also, Boyle and Russell [15] studied endwall heat transfer on a six-vane linear cascade for mainstream Reynolds numbers between 73000 and 495000 which is based on the inlet velocity and the axial chord. The authors found the heat transfer distribution on the endwall of a turbine stator vane being strongly influenced by the flow through the cascade. At low Reynolds numbers the heat transfer contours are aligned with the inviscid flow streamlines. It differs significantly for high Reynolds numbers and correlates with the local freestream velocity.

Wedlake et al. [133] conducted heat transfer experiments on an annular cascade of nozzle guide vanes in an isentropic light-piston cascade. Measurements of local heat-transfer rates are carried out around the blade surface and on the endwalls. Endwall heat transfer and pressure measurements did not indicate enhanced heat transfer being caused by crossflow vortices. Takeishi et al. [124] also studied vane surface and endwall heat transfer in an annular vane cascade but with low aspect-ratio turbine nozzles of 0.5. Increased heat transfer and decreased film effectiveness was found near the leading edge due to the present horse-shoe vortex. Small influence on heat transfer and film effectiveness was stated for the pressure side of the vane. The film effectiveness results are presented versus a normalized distance $x/M\,S_{eff}$ along potential streamlines where S_{eff} is the equivalent film-cooling slot width.

2.2 Film-Cooling Situation

2.2.1 Aerodynamic Conditions

As shown in the last section, the turbine components exposed to the hot gases are extensively cooled to a large part by film-cooling. The coolant is transported through small holes which are drilled into the component walls. It leaves the holes to form discrete jets which are injected into the wall boundary layer. Due to lateral and radial pressure gradients in a turbine blading, the near-wall region of the platforms is dominated by secondary flow pattern. This flow field and its turbulence promotes the streamwise transport and diffusion of the film-cooling jets. Due to viscous interaction of the hot gas path and the coolant with the adjacent walls, a complex three-dimensional jet flow field is created which is characterized by a system of vortex structures. Furthermore shear layers with increased turbulence develop in the region of the vortices due to velocity differences between the coolant and the hot gas (Baldauf [4]). Strongly pronounced vortical structures can be seen for wall-normal injection where the topology of the flow structure is described by Kamotani and Greber [64] or Andreopoulos and Rodi [1,2]. Margason [92] gives an extensive overview on the research of a jet in cross flow. The cross-section of the exiting transverse jet is dominated by the formation of a counter-rotating vortex pair studied by many researchers as Blanchard et al. [11], Cortelezzi and Karagozian [27], Smith and Mungal [122] or Fric and Roshko [35].

The system of vortices is not only present in case of transverse jets. It can also be observed for injection through inclined holes though in a diluted form. Extensive numerical studies from the group of Leylek [19,57,93,131] were presented in a four-part paper investigating different film-cooling configurations (straight cylindrical holes, compound angle holes, fan-shaped holes with stream-wise and compound-angle injection). The dominant physical mechanisms were described that are responsible for the individual cooling performance. The numerical simulations done by Walters and Leylek [131] identified the flow field at the hole entrance to be very complex and depending on the operating conditions. The complexity is mainly caused by the coolant flow-path into the hole. Here, the coolant is strongly accelerated being channeled from a large volume into the small film hole. Furthermore, the coolant experiences excessive turning at the sharp-edged entrance which results in a three-dimensional separation region on the downstream surface of the hole. This separation region is the main cause for a non-uniform coolant distribution and increased turbulence inside the hole. According to Walters and Leylek [131] the counter-rotating vortex pair downstream of the jet exit is the most significant mechanism that affects the cooling performance. Its origin can be associated with the vorticity inside the film hole.

The complicated flow field of a film-cooling jet that enters the main flow gas path is displayed in Fig. 2.5. The most obvious effect is the gradual jet deflection toward the wall surface. The film jet is bent over by the cross flow which is laterally and vertically displaced showing increased streamwise velocities in the upper shear layer. Furthermore, it wraps around the film jet acting

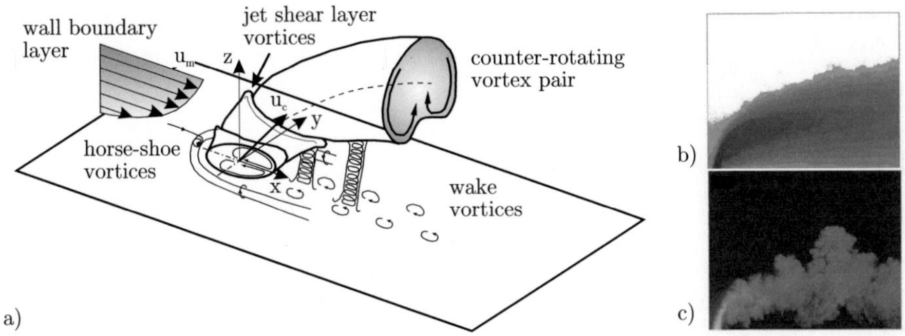

Figure 2.5: a) Flow topology and coherent vortex structures for a film jet in cross flow according to Fric and Roshko [35] and (b-c) ensemble-averaged (top) and instantaneous concentration image (bottom) of a transverse jet in cross flow ($u_c/u_m = 5$, $Re_c = 16600$) obtained by Smith and Mungal [122] using Planar Laser-Induced Fluorescence (PLIF)

as a rigid obstacle. In case of low coolant momentum, the main flow acts as partial cover causing the film trajectories to start bending early inside the film hole. The film is turned around the trailing edge of the hole to attach to the wall.

Increasing the coolant momentum causes the exiting jet to be less affected near the exit and to penetrate further into the cross stream. Depending on the coolant momentum the film jet is bent back to reattach to the wall. Near the holes, the film jet partially detaches from the wall with increasing coolant momentum. A small separation region with reverse flow occurs which leads to hot gas entrainment. The hot gas fluid travels upstream and is lifted upwards by the shear layers at the film edges to be transported downstream along with the jet fluid. For any ratio of coolant-to-main flow momentum, the wake region downstream of the film jet is characterized by a complex three-dimensional vortical movement. It mainly consists of four coherent vortical structures following the classification proposed by Fric and Roshko [35]. These are turbulent structures that possess coherent eddies of different time scales and sizes. The majority of them arises from vortical interaction within the jet shear layer in the near-field of the hole.

Counter-Rotating Vortex Pair (CVP). The wake flow field is dominated by two counter-rotating vortices which still can be observed far downstream of the hole exit. After leaving the hole, the jet is bent toward the wall by the cross flow. This is due to pressure forces between the luv-side (jet upper side) and lee-side (jet bottom side) of the jet and due to viscous forces inside the jet shear layers. The pressure difference between those two sides due to its bending causes a downward, rotational movement at the jet edges which is continued back upwards inside the jet core. At the hole entrance, the coolant experience excessive turning which causes pressure differences between the downstream (increased velocity, low pressure) and upstream hole edge (decreased velocity, high pressure). This process drives a downward transport of fluid inside the wall boundary layers and back toward the hole center forming a counter-rotating vortex pair. For inclined jets, the coolant turning at the hole entrance can be so large that the secondary flow vortices persist across the entire hole length and emanate into the cross flow. This further amplifies the rotational movement initiated by the jet bending. Combined by these two mechanisms, the two separate vortices with opposite sense of rotation (commonly known as CVP) are convected downstream. Their sense of rotation promotes jet lift-off and hot gas entrainment underneath the jet core. Due to the specific shape of the streamwise jet cross-section, their name "kidney-vortices" is derived. Experiments by Smith and Mungal [122] suggest that, under specific circumstanced the instantaneous flow field of the CVP can either be symmetric or asymmetric in shape. Thus the jet in the far-field can reveal axisymmetric or sinusoidal motion of the CVP. Yuan et al. [141] conclude from LES-computations that the position and strength of the instantaneous CVP varies with time and streamwise location. The numerical simulations further show that the magnitude of the vorticity for the CVP in the interacting flow field persists longer and decays at much smaller rates than the fluctuating vorticity in the rest of the flow.

Jet Shear-Layer Vortices. The jet shear-layer vortices consist of ring vortices which are formed by Kelvin-Helmholtz instabilities in the mixing layer between the jet and the cross flow. Such effects occur in free shear-layers formed by the merge of two flows with a sufficient velocity difference across the interface between the two fluids. The Kelvin-Helmholtz instability describes the propagation of an external disturbance impressed on the shear layer in time and space. It depends on the resulting flow profile in the free shear-layer which exhibits an inflexion point and can therefore cause unstable flow conditions. Small perturbations may give an oscillation of the vortex sheet with its amplitude growing considerably unstable.

Horse-Shoe Vortices. The horse-shoe vortices form upstream of the film exit in the stagnation region of the cross flow boundary layer. The injected film jet acts as a soft obstacle showing nearly the form of a cylinder. The upstream boundary layer suffers strong deceleration which induces local flow separation. Encountering an adverse pressure gradient, the near-wall portions of the boundary layer begin to roll-up into spiral vortices which are separated into two different branches. The vortex strength is weaker compared to the vortex structure induced by a rigid cylinder. The reason for this effect are the additional shear forces that occur in the shear layer between the jet and the stagnation line of the cross flow. The legs extend around the jet column and are convected in streamwise direction.

Wake Vortices. Two vortical structures can be observed in the wake of the jet: streamwise vortices which lie close to the wall and vertically oriented (cyclonical) vortices which extend from the jet body down to the wall. These unsteady vertical vortices are generated from the vorticity of the cross-flow boundary layer. It arises directly from the interaction of the horse-shoe vortex and the vertical jet flow (Yuan et al. [141]). According to Fric and Roshko [34] or Kelso et al. [68] separation events on the lee-side of the jet column cause the formation of small vortices. They turn up and become vertical, cyclonical wake structures. These structures are shed off from the exiting jet with its downstream progression reminiscent of a Karman vortex street induced behind a solid cylinder in cross flow. For inclined jets, the Strouhal number based on the jet diameter matches closely the value obtained for a cylinder with the same diameter (see Moussa et al. [98]).

Since the velocity profile of the emanating film jet is linked to the flow field inside the hole, the flow conditions at its inlet can significantly influence the jet in cross-flow interaction (see Burd and Simon [22]). Different factors cause a three-dimensional, non-uniform flow field at the hole entrance that proceeds into the hole. Examples are the ribbed walls of the internal cooling channels, the position of the ribs with respect to the film hole, impingement jets or the orientation of the cooling channels.

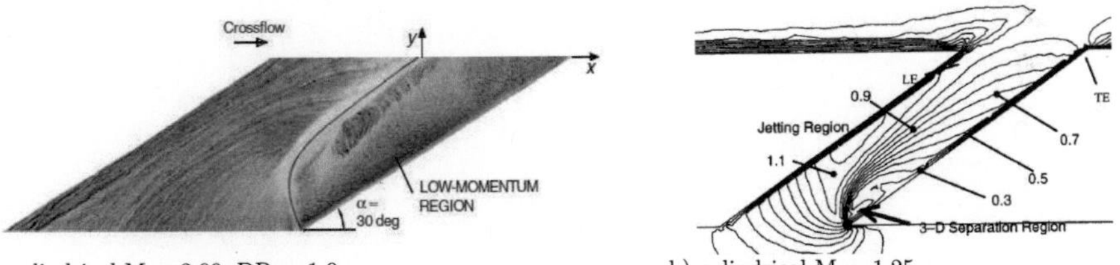

a) cylindrical M = 2.00, DR = 1.0 b) cylindrical M = 1.25

Figure 2.6: a) Oil-film flow visualization in a film-cooling hole (blowing rate $M = 2.0$) inclined at 30° done by Lee et al. [77] and b) velocity distribution for a cylindrical hole (blowing rate $M = 1.25$) inclined at 35° computed by McGovern and Leylek [93]

For inclined holes, the coolant experiences strong deflection when it enters the hole and the streamlines show distinct curvature. As a result of this, the coolant separates from the downstream wall to form a zone with detached flow that can spread far into the hole (see also Fig. 2.6). It influences the flow characteristics inside the hole and the heat transfer. The extent of the separation region toward the hole center reduces the effective hole cross section. It forms a "vena contracta" that results into increased velocities and distorted, asymmetric velocity profiles at the hole exit. This effect is commonly denoted as jetting effect. By using short holes, this effect can increase the effective injection angle and therefore lead to earlier jet detachment from the surface (as observed by Lutum and Johnson [90]). The formation and strength of these effects mainly depends on the geometrical and fluid parameters of the holes like inclination angle, hole length, blowing parameter and the orientation of the cooling channel. Due to the streamline curvature at the hole inlet, the fluid experiences an decrease in velocity toward the downstream wall while the flow velocity increases on the upstream wall. That results in a pressure gradient across the hole section driving a cross flow inside the wall boundary layer to form the two branches of the counter-rotating vortex pair.

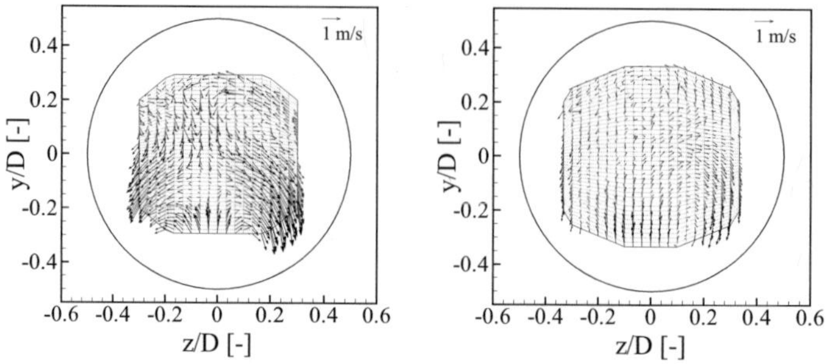

(a) secondary flow velocity field inside a cylindrical film-cooling hole inclined at 42° for M = 0.5, x_h/L_h = -0.46 (left) and x_h/L_h = -0.21 (right)

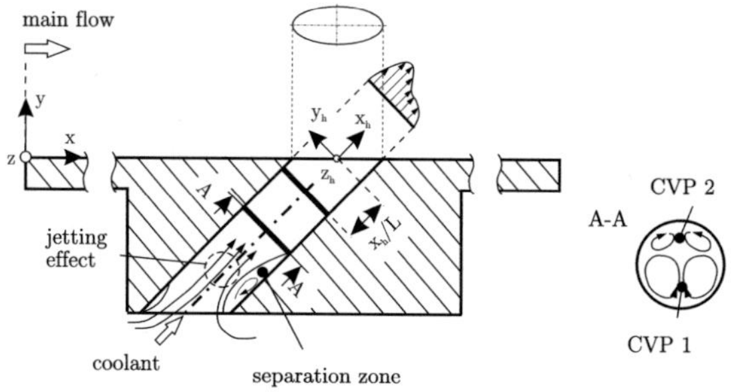

(b) schematic of flow structure in a cylindrical hole with location of measurement planes

Figure 2.7: Flow field inside a scaled-up film-cooling hole obtained by Laser-Doppler-Anemometry (LDA) for a blowing rate of $M = 0.5$ and two different positions x_h/L_h with regard to the hole exit by Kunze and Vogeler [74]

Flow field investigations were performed by Kunze and Vogeler [74] on a single, scaled-up film-cooling hole with a diameter $D = 15\,mm$ and an inclination angle $\alpha = 42\,°$ using Laser-Doppler-Anemometry. To obtain the vortical velocity components directly, the 2D - optical probe must be aligned with the hole axis. It is traversed in orthogonal measurement planes containing over 700 grid points. Resolution of the hole cross section was limited to about 60 % to avoid reflection and disturbances of the laser beams at the hole walls. The effective measurement section is reduced at greater depth of the plane. The planes are positioned with regard to the center of the hole exit at a distance of 17.5 and 37.5 mm. That translates into a non-dimensional distance of x_h/L_h = -0.21 and -0.46 with regard to the hole length. Two different blowing ratios of M = 0.5 and 1.0 were investigated. The density ratio is unity using air as fluid for main flow and coolant at isothermal conditions. The results of the velocity field inside the hole (Fig. 2.7 a) show the counter-rotating vortex pair and its development along the hole axis. It can be observed that the strength of the vortical motion is larger at smaller distances from the hole inlet. Two strong counter-rotating vortices are observed located at the downstream wall of the hole, denoted as CVP 1 in Fig. 2.7 b. Although an increase in coolant mass rate and blowing rate results in an enhancement of the vortex pair, the strength of the vortical motion is reduced along the hole axis due to mixing processes. Each branch of the CVP 1 drives another but significantly weaker vortex (CVP 2) that develops at the lee-ward side of the hole rotating in opposite direction.

2.2.2 Thermal Conditions

Film-cooling injection as method to cool turbine components introduces a layer of fluid inside wall boundary layer that has lower temperatures compared to the main flow. That mainly affects the fluid temperatures near and at the wall. To characterize the heat flow from the injected film layer into the wall, the change in the temperature profiles and the surface temperature must be considered. Typically, the temperature gradients parallel to the wall are assumed to be negligible small compared to the gradient normal to surface. Therefore, the axial and lateral components of the heat flux vector are neglected so that the three-dimensional heat transfer situation is reduced to a one-dimensional problem.

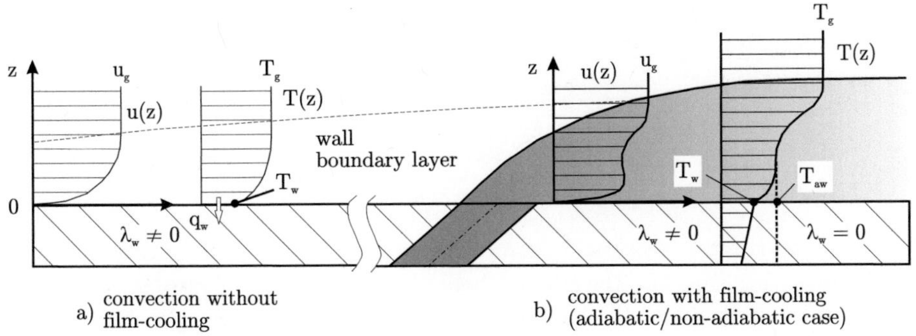

Figure 2.8: Schematic of temperature profiles on a cooled surface and influence of film-cooling on the main flow field and heat transfer

A schematic of the temperature profiles for the cases with and without film-cooling is displayed in Fig. 2.8. It further shows the influence of the injected cooling film onto the local near-wall flow field. Far off the injection location, the convective heat transfer is driven by the thermal boundary layer that develops due to the temperature difference between main flow and solid wall. The fluid temperature decreases toward the solid surface and finally approaches the wall temperature. The heat transport inside the wall boundary layer and into the wall is modeled by the concept of the heat transfer coefficient. This parameter is influenced by the local flow field conditions, hence the fluid and the wall temperature as well as its physical properties.

The heat transfer coefficient for a turbulent flow over a flat surface without film-cooling injection relates the driving temperature difference to the convective heat flux into the wall as

$$\dot{q}_w = h_0 \left(T_{rec,g} - T_w\right) \tag{2.1}$$

Hence the convection problem can be described by only the temperatures of the main flow and at the wall. Applying film-cooling to the surface adds the additional temperature of the injected coolant to the problem. This extends the convection situation to a three-temperature problem. The injection of coolant onto the surface with a temperature lower than the main flow temperature significantly alters the near-wall flow field and the approaching wall boundary layer. In addition, it considerably influences the thermal and fluid boundary conditions for the heat transfer process. Since the film is injected at velocity magnitudes different from the main flow velocity, it changes the velocity gradient at the wall which also affects the convective heat flux from the film into the wall. Outside of the film, the temperature profile $T(z)$ is equal to the main flow temperature T_g. Within the induced film-cooling layer, the driving temperature gradient for the heat transfer is steadily decreased toward the film temperature T_f at the wall. In a case of adiabatic conditions for the solid wall, which means that either the wall is perfectly insulated or the wall material has zero heat conductivity ($\lambda = 0$), the local film temperature at the wall equals the adiabatic wall temperature T_{aw}. Based on equation (2.1), the convective heat flux with film-cooling along the cooled surface can be expressed as

$$\dot{q}_f = h_f \left(T_{aw} - T_w\right) \tag{2.2}$$

where h_f is the heat transfer coefficient due to the presence of film-cooling and (T_{aw} - T_w) is the driving wall temperature difference comparing the adiabatic case with the one implying heat conduction through the wall. The adiabatic wall temperature is the effective gas temperature at the wall and depends on the temperatures of the main flow and the injected coolant gas. It is further influenced by the mixing process of the film-cooling jets. The resulting surface temperature distribution is described by the non-dimensional wall temperature for an adiabatic case, commonly denoted as the adiabatic film effectiveness

$$\eta_{aw} = \frac{T_{rec,g} - T_{aw}}{T_{rec,g} - T_c} \tag{2.3}$$

where $T_{rec,g}$ is the recovery temperature $T_{rec,g}$ and T_c is the coolant temperature at the injection location. On the hot gas path side, the flow conditions are of compressible character due to high velocities. In compressible flows, viscous dissipation within the boundary layer causes the near-wall fluid temperature to increase. To account for this effect, the recovery temperature $T_{rec,g}$ is expressed with the main flow temperature T_g according to Schlichting [114] as

$$T_{rec,g} = T_g \left(1 + r\frac{\kappa - 1}{2} Ma_g^2\right). \tag{2.4}$$

The parameter r is the recovery factor which is defined with $r = Pr^{1/2}$ for laminar boundary layers and $r = Pr^{1/3}$ for turbulent boundary layers. For incompressible flows ($Ma_g < 0.3$), the dissipative temperature increase is negligible so that the recovery temperature $T_{rec,g}$ can be replaced by the static temperature T_g. This leads then to the definition of the adiabatic film effectiveness as

$$\eta_{aw} = \frac{T_g - T_{aw}}{T_g - T_c}. \tag{2.5}$$

The film effectiveness in a non-adiabatic case that is derived from surface temperature measurements is defined for compressible flows as

$$\eta_f = \frac{T_{rec,g} - T_w}{T_{rec,g} - T_{pl}}. \tag{2.6}$$

The wall temperature T_w incorporates heat conduction and possible radiation effects which mostly occur in high-temperature rigs. Due to the limited access in experimental setups, the coolant temperature is difficult to obtain without any disturbance of the jet flow field. Therefore, the plenum temperature is used instead which neglects the heat pick-up or heat loss inside the holes. Dependent on the used temperature ratio, it leads to an increase or decrease in coolant exit temperature. Restricting the flow field conditions to low Mach numbers ($Ma_g < 0.3$) and incompressible flow conditions, typically two assumptions can be made. At first, the dissipative temperature increase is neglected so that the recovery temperature can be replaced by the hot gas temperature. Second, assuming the density change across a turbine airfoil in a low speed test rig that operates at inlet Mach numbers below 0.1 to be small, the hot gas temperature T_g can be replaced by the static main flow temperature T_m at the cascade inlet.

To obtain the film effectiveness for an adiabatic case, the effects of heat conduction on the wall temperature T_w have to be eliminated from the measured surface temperature fields. One approach is a thermal analysis based on a Finite-Element method of the solid test body. In addition, this method can be used to evaluate the heat pick-up or heat loss by a simple one-dimensional heat balance across the cooling holes. The computed conductive heat flux pattern is then superimposed onto the measured temperature fields to reduce the heat conduction effects. The adiabatic film effectiveness for an inverse coolant-to-main flow temperature ratio and incompressible flows is then computed as

$$\eta_{aw} = \frac{T_{aw} - T_m}{T_c - T_m} \tag{2.7}$$

with T_{aw} the adiabatic wall temperature, T_c the coolant temperature at the hole exit and T_m the static main flow temperature which is back-calculated from the total temperature measured by a thermocouple at the cascade inlet.

The definition in equation (2.5) relates the actual temperature difference between the adiabatic wall temperature and the surface temperature of the solid wall to the possible temperature reduction that can be achieved with cooling. This is the case when the adiabatic wall temperature approaches the coolant temperature T_c of the injected film. Then the adiabatic film effectiveness reaches its maximum of unity. Substituting (2.5) into (2.2), one can deduce the convective heat flux with film-cooling in dependency of the adiabatic film effectiveness as

$$\dot{q}_f = h_f \left[(T_g - T_w) - \eta_{aw}(T_g - T_c) \right] \tag{2.8}$$

Taking into account the preceding conclusions, cooling performance with film-cooling can not be evaluated on the basis of film-cooling effectiveness or heat transfer coefficient alone. Instead, both quantities must be equally considered since the film injection involves adverse effects due to these parameters. It reduces the fluid and wall temperature but also increases the heat transfer by adding additional turbulent structures inside the jet core and the shear layers at the outer film bounds. The change in heat load due to film-cooling can be determined by combining the heat flux without film-cooling (2.8) and the local heat flux in presence of film-cooling (2.1) which then defines the local heat flux ratio $\dot{q}_f/\dot{q}_w$ as

$$\frac{\dot{q}_f}{\dot{q}_w} = \frac{h_f}{h_0} \left(1 - \eta_{aw}\theta_w \right). \tag{2.9}$$

Here θ_w is the non-dimensional wall temperature given by

$$\theta_w = \frac{(T_g - T_c)}{(T_g - T_w)}. \tag{2.10}$$

Its reciprocal can be described as overall cooling effectiveness $\theta_w = 1/\eta_w$ with film-cooling injection in case of a solid, non-adiabatic wall.

2.3 Influencing Parameters on Film-Cooling

Discrete film-cooling is an efficient way to protect the surface of hot turbine components from failure and keep the metal-surface temperatures below the limits dictated by the material. The process of film-cooling has been investigated over the past centuries not only for various cooling geometries like cylindrical, fan-shaped, laid-back fan-shaped or conical holes, but also for a variety of main flow and coolant flow parameters. The hole geometry itself has an tremendous effect on the cooling performance. Hence, many investigations were focussed on the effect of various geometry features on the film-cooling effectiveness and the heat transfer.

A large data base is available in literature containing measurements with compound angled holes, various inclination angles, pitch-to-diameter ratio, spacing ratio, diffuser angles and so on. Here, film effectiveness and heat transfer data was acquired for different blowing and density ratios. Since published data is vast and the testing conditions differ significantly between the experimental setups, only a brief overview on the findings can be given here to show the trends of the influencing parameters on film-cooling. Hence, the following study represents only a small part of the whole picture and does not claim for completeness. Subsequently, the influencing parameters on the film-cooling process are listed in table 2.1 as proposed by Kissel [69]:

	geometrical parameter	aerodynamical parameter
primary	hole shape	blowing ratio
	hole length	density ratio
	hole pitch & stagger	momentum ratio
	hole angles (axial, lateral)	
	diffuser length	
	diffuser angles (axial, lateral)	
secondary	hole location on platform/airfoil	unsteady effects
	wall curvature	(rotor-stator interaction)
	surface roughness	rotational effects (blades)
	coolant supply:	**effects of main flow:**
	rib geometry	turbulence intensity
	rib pitch/inclination	Mach number
	orientation & shape of	boundary layer thickness
	cooling channels	acceleration
		pressure gradient
		(secondary flows)

Table 2.1: Influencing parameter of film-cooling

Comprehensive description of the parameter of influence for the film-cooling process can be found in, e.g. Bogard and Thole [12], Kissel [69] or Lutum [89]. For further details on the technological background on shaped holes, the reader is referred to Bunker [21].

Basically, film-cooling is a mixing process of two fluids with different flow field structures and temperatures where the velocity and turbulence distribution of both gases have significant influence on the resulting cooling efficiency of the cooling films. For film-cooling, air is drawn from the last compressor stages to be guided through the secondary air system toward the turbine components. To initiate the injection process through the holes drilled into the component walls, the pressure losses associated with the coolant transport must be lower than those across the combustion chamber and the hot gas path up to the injection location. This can be characterized by the coolant-to-main flow pressure ratio $\Pi_c = p_{tc}/p_m$. For pressure ratios greater than one, cooling fluid flows through the hole to be injected onto the outer wall surface. Considering the fluid properties for main flow and coolant, the pressure ratio can be rearranged to represent the coolant-to-main mass-flow ratio. In this case, an increase in coolant pressure means also an increase in cooling mass flow rates. Due to friction losses inside the cooling channels, the decreasing coolant supply pressure in streamwise direction results in lowered hole mass flow rates from hole to hole. Applying an improved characterization of the injection process for a given hole geometry, the blowing ratio is one of the key parameters for film cooling. It considers the ratio of mass flow rates of coolant and the main flow and is defined as

$$M = \frac{\rho_c u_c}{\rho_m u_m} = \frac{\frac{\dot{m}_c}{A_c}}{\frac{\dot{m}_m}{A_m}}. \tag{2.11}$$

The expression of ρu can be rewritten to the corresponding mass flow rate normalized with the cross-sectional area, usually denoted as mass flux. An increase of the blowing ratio results into higher coolant mass flow rates enhancing the cooling performance on the component's wall surface. However, the injected film-cooling jets tend to lift-off from the surface at higher blowing rates depending on the used geometry (cylindrical or fan-shaped) and inclination angles of the holes. In general, the film efficiency continually increases with enhanced blowing rates as long as the cooling film stays attached to the wall. The optimum blowing rate to reach maximum film efficiency is significantly dependent on all other influencing parameters and is difficult to predict. But some generalized trends can be derived from the previous research investigations.

Using plain cylindrical holes, the maximum in film-cooling efficiency is reached at blowing rates of approximately 0.5 (Goldstein et al. [42]). Beyond that, it significantly drops down to very low levels which is due to the jet lift-off from the surface. This effect is initially limited to the near-injection region of the hole. It shows a local jet lift-off at the sharp turning around the downstream hole edge with reattachment farther downstream. As the blowing rate is further increased, it leads to complete wall detachment and penetration far into the cross flow. Although shaped holes are also designed with an cylindrical inlet section, the diffuser-shaped hole exit has a tremendous effect on the spatial coolant distribution across the wall. The film jet experiences a reduction in velocity and momentum as well as an enhanced lateral spreading and flattening in wall-normal direction. That in turn results in a decreased tendency of lift-off. Therefore, the maximum of film-cooling effectiveness is about $20 - 30\,\%$ higher compared to cylindrical holes. Also, the corresponding blowing rate is found to be larger of about 1.5.

According to Lutum [89], three major flow regimes can be identified for the film-cooling. At small blowing rates the film is deflected toward the wall by the main flow due to its low momentum and keeps attached while the blowing rate increases. Here, the film effectiveness correlates with the blowing ratio, shown by Pedersen et al. [102]. With further increase of the coolant-to-main-flow mass flux ratio up to moderate values, the film effectiveness reaches its optimum with more pronounced lateral and axial wall coverage of the films. At large blowing rates, the cooling film tends to partial or complete detachment from the wall due to its high momentum. This leads to a dramatic decrease in film effectiveness which does not correlate with the blowing rate anymore. The results from Pedersen et al. [102] showed that the coolant-to-main flow density ratio

$$D_R = \frac{\rho_c}{\rho_m} \tag{2.12}$$

has a tremendous effect on the film effectiveness at greater mass flux ratios. Assuming the same type of gas used for coolant and main flow, the density ratio is mainly influenced by the temperature ratio T_R. This implies that the tendency of the film lift-off from the surface is indirectly influenced by the density ratio by means of the momentum flux ratio, defined as follows

$$I = \frac{\rho_c u_c^2}{\rho_m u_m^2} = M^2 \frac{1}{D_R}. \tag{2.13}$$

From (2.13), it is clearly visible that the jet momentum and the tendency of jet blow-off is reduced with increasing density ratio while main flow conditions and coolant mass flow rate remain unchanged. Consequently, the optimal blowing rate, where maximum film-cooling effectiveness can be achieved with a given hole geometry, is shifted toward greater magnitudes. With the selection of two out of the three given film blowing parameters (blowing rate M, momentum flux ratio I and density ratio D_R), the third component is defined by the other two and can not be changed independently.

2.3.1 Effect of Cooling Hole Geometry

Hole Shape One of the goals in applying discrete film-cooling is to establish ideal cooling films as those initiated from two-dimensional continuous slots. Due to many constraints (aerodynamics, mechanical stresses, thermal loads), it is not preferable to place large discontinuities in high-temperature turbine components weakening the mechanical structure. As a consequence, the film-cooling process is applied almost exclusively through discrete holes and evenly spaced rows. For more than three decades, cylindrical holes have been widely for cooling purposes. Research on that topic was done on essentially every aspect of film jet and main-flow interaction containing film-cooling effectiveness and/or heat transfer coefficient as spatial distribution or laterally-averaged quantities. So the literature on this hole geometry is quite vast, citing here only some examples as Goldstein et al. [42], Lutum and Johnson [90] or Yuen et al. [142].

The film-cooling jet initiated from a cylindrical hole features a complex three-dimensional flow field with vortical movement inside the jet core and in the interaction zone between jet and main-flow field. Although the holes are inclined at shallow angles, wall attachment of the cooling films remains only for a small range of blowing rates. Hence the injection process is significantly dependent on the blowing ratio. The cooling performance reaches its maximal film effectiveness at quite low coolant mass flow rates. At greater blowing rates above 0.5, the jets tend to lift-off which results into a significant drop in film effectiveness. Due to the vortical flow field (see also section 2.2.1), partially hot gas fluid is transported toward the wall surface to reduce the cooling performance. It further introduces additional turbulence and enhances the heat transfer.

To enhance the operating stability of cooling films and to reduce the tendency of jet lift-off, the hole exit geometry was changed from round to a diffuser-like shape. Shaped holes are composed of a round metering or throat section over some portion of the hole length. It is followed by an expanded uniform and symmetric exit region on the cooled hot gas surface. The shaping is commonly confided to the last part of the hole over two or three hole diameters with divergence angles of 10° to 15° in lateral and axial direction. It intends to decrease the coolant momentum and to produce a flatten jet shape with reduced jet core turbulence and hence decreased heat transfer enhancement. Because the exit area is increased by the factor of two to three, this leads to significantly enhanced lateral diffusion of the coolant without separation.

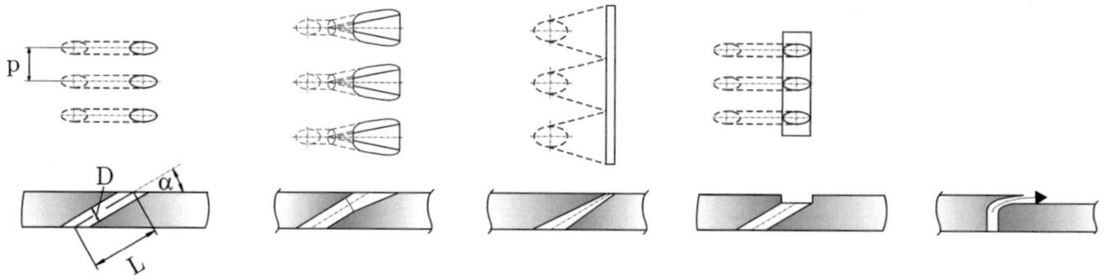

a) cylindrical holes b) shaped holes c) console holes d) transverse slot d) continous slot

Figure 2.9: Different hole geometries for film-cooling technology

The introduction of diffuser-shaped hole exits leads to a reduction of coolant momentum, less aerodynamic mixing losses, improved lateral coolant coverage and higher film effectiveness. Also less influence was observed on hot gas flow parameters as Mach number, turbulence intensity or boundary layer thickness (Saumweber et al. [112] or Gritsch et al. [48]). A variety of different shapes of diffuser holes exist in literature. Bunker [21] tried to classify them a methodology proposed in his detailed technological review on shaped holes. Here the fan-shaped hole geometry is distinguished between lateral or axial diffusion or a combination of both.

Besides, alternative hole shapes have been investigated (see Fig. 2.9) and some interesting remarks are stated in [126]. Sargison et al. [108, 109] studied a converging slot geometry, denoted as console hole. Here, the cross section starts from nearly cylindrical at the hole inlet and changes to slot-like shape at the hole exit. Due to that, the hole area is decreased, so that the minimum hole area and the highest velocities are located at the hole exit. It further applies lateral expansion of the jet to approach near-slot injection characteristics. The design of the console hole intends to decrease the aerodynamic losses inside the hole by forcing the wall boundary layer into a laminar state. The low turbulence at the hole exit and the resulting stable coolant layer should reduce the interaction and mixing with the hot gases. The results indicate that the thermal performance in terms of heat transfer coefficient and film effectiveness is similar to what can be achieved with classical hole shapes. In contrast, the aerodynamic loss of the console hole is lower. Cascade tests of the console geometry on the rear part of the vane suction side revealed higher effectiveness but also enhanced heat transfer.

Kohli and Bogard [71] modified the airfoil suction side contour near the leading edge to establish a type of Coanda-effect for the exiting film jets. It was intended to keep the films attached to the surface with improvement of the local cooling performance. The case at high blowing rate shows nearly doubling of the film effectiveness close to the hole exits. The results of this study exhibit a quite large potential for improving the cooling process but still attention must be drawn to effects on the aerodynamic flow field and loss issues.

More recently, small two-dimensional surface slots had been introduced at the hole exits which are denoted as trenches. The slots are fed mostly by cylindrical holes. Bunker [20] investigated the influence of a continuous trench with two width ratios of $W/D = 1.13$ and 1.5 and slot depth of $S/D = 3$ on the adiabatic film-cooling effectiveness for a row of cylindrical holes. The results show that for a wider slot, the film effectiveness improves with increasing blowing rates but still exhibits some sensitivity below $M = 2$. According to the author, this may indicate hot gas entrainment into the trench which reduces the lateral spread of film jet within the slot. Decreasing the slot width, the film effectiveness reveals a remarkable insensitivity over the blowing ratio and the results showed nearly the same cooling performance for all test cases. Hence, injection into a narrow slot gives the highest film effectiveness for low blowing rates compared to a wider trench and the non-slot holes.

The author further investigated a shallow hole-in-slot configuration with a depth of less than a half hole diameter. In comparison to conventional axial film holes, the shallow trench experience provides better cooling performance at elevated film effectiveness about 50 to 75 % in the near-hole region ($x/D \leq 40$). Due to its small depth, a shallow trench can either be incorporated into the protective coating or within the substrate material with less intrusion than deeper slots. The author proposed that this improved performance might be due to the trench forming a subsurface that prevents interaction of the exiting jets with the hot gases. The downstream edge acts as a barrier to force some portion of the exiting coolant to spread in lateral direction. It further disturbs the typical vortex structure of a film-jet in cross flow to sweep hot gas underneath the films.

Also Lu et al. [88] studied different trench configurations applied to a row of cylindrical holes. The results showed increased film effectiveness but also enhanced heat transfer coefficients. Further related work on hole-in-slot film-cooling configurations can be found in [52, 85–87]. One of the approach's disadvantages is the weakening of the structural integrity of the blades and platforms by locally reducing the material thickness. It further induces small notches that would allow stress peaks to occur with the result of cracking issues and reduced lifetime.

Hole length-to-diameter ratio L/D The length of a film-cooling hole dictates how and to what extent the flow field effects at the hole inlet are transported through the hole to affect the exit velocity distribution and the coolant mixing process with the hot gas. Lutum et al. [90] measured film effectiveness for a row of cylindrical holes over a broad range of hole-length-to-diameter ratios ($L/D = 1.75 \ldots 18$) and blowing rates ($0.5 < M < 1.56$) to systematically investigate its influence on the film-cooling process. In the range of $1.75 < L/D < 5$ the film effectiveness decreases with reduced hole length ratios L/D which is attributed to the undeveloped flow inside the cooling channels. Beyond a length-to-diameter ratio L/D of 5, no changes in the film effectiveness was found for larger L/D. As stated by the authors, this effect results from the sufficiently developed flow characteristics inside the hole. Also Wilfert and Wolff [139] studied two different hole length ratios at $L/D = 4$ and 8 but in the context of studying internal flow conditions and plenum geometry on film effectiveness.

In general, the reduction of the hole length below certain limits of $L/D = 5$ leads to significantly decreased film effectiveness. This effect is caused by the flow field inside the cooling holes. At small hole length ratios, the separation zone that occurs on the downstream hole wall affects the flow conditions at the hole exit in several ways. Due to the sharp turning of the cooling flow, counter-rotating vortices develop in a transverse plane inside the hole and are transported toward the hole exit. This leads to considerable turbulence production inside the hole and concurrently to enhanced turbulence in the jet core. As the vortex system exits the hole, the intense rotational movement increases the mixing process of the films with the main flow by augmenting the transport of hot gas fluid toward the cooled surface. Furthermore, the radial spread of the separation zone significantly reduces the effective hole cross-sectional area which leads to increased coolant velocities in that region (denoted as jetting effect). At small hole length ratios of $L/D \sim 2$, the separation zone and the strong secondary vortices can spread into the hole exit region which drastically alters the velocity distribution and the turbulence inside the jet core. In this case, the jetting effect inside the hole attributes also to increased effective injection angles. In combination with increased jet momentum, the tendency of film detachment off the cooled wall is enhanced.

Inclination angle α Mostly in real engine applications and scientific research setups, the cooling hole is designed with a quite shallow inclination angle between 30° and 35° with regard to the main flow direction. Exceptions are the intersection regions between the airfoil and the platforms. This is mainly due to alignment problems for the manufacturing tools to produce such shallow holes. Therefore film-cooling holes in those regions often feature more steep inclination angles of up to 40°. This has in turn decisive influence on the tendency to lift-off with greater jet momentum.

During the injection process, the inclination angle dictates the ratio between the tangential and wall-normal vector component of the jet momentum. With shallow injection, the wall-normal momentum component is reduced to produce less tendency to lift-off while exiting from the hole in tangential direction. Since the wall-normal momentum component is increased with larger inclination angles, the film jet tends to detach earlier from the wall. This happens also at lower blowing rates due to the increased turning of the cooling flow at the hole exit location. Film injection with a very steep inclination angle of $\alpha = 90°$ can benefit the film distribution in streamwise direction due to the very sharp turning of the jet. Foster and Lampard [33] have observed elevated heat transfer coefficients directly downstream of the hole exit for wall-normal injection. The augmentation of heat transfer rapidly decreases in streamwise direction. The heat transfer coefficient for shallow injection at $\alpha = 35°$ is significantly higher in the hole far-field due to the improved film integrity and reduced mixing.

Compound angle β Experimental investigations on platforms of cooled turbine blades, as conducted by Friedrichs [39], Vogel [127] or also Barigozzi et al. [6], have clearly pointed out the significance and the influence of the near-wall cross flow on film-cooling injection. Studies on flat plate film-cooling done by Schmidt et al. [117], Ekkad et al. [32], Kaszeta [66] or Ligrani [80] showed the influence of the three-dimensional cross flow on cooling performance by introducing laterally inclined holes. Here, the varied parameter is the angle between the hole axis and streamwise direction (denoted as compound angle). Typical values of the compound angle β can range from 0° (in line with main flow), 45°, 60° up to 90° (normal to main flow).

In this context, additional lateral hole inclination can produce an increase of lateral coolant spread, hence more pronounced wall coverage. Ekkad et al. [31] have investigated a row of five cylindrical holes with a compound angle of $\beta = 0°$, 45° and 90°. The results showed that in all cases, the coolant injection with a lateral compound angle leads to an increase in laterally-averaged film effectiveness. In case of the obtained heat transfer, the application of a laterally-angled hole results in enhanced jet core turbulence, hence in a significant increase of the heat transfer coefficient. Also Schmidt et al. [117] observed increased laterally-averaged film effectiveness for all momentum flux ratios up to $I = 4.0$. The experiments were conducted on cylindrical holes with a compound angle of 0° and 60°, respectively.

Brauckmann et al. [18] experimentally investigated the influence of the compound angle at specific values of 0°, 30° and 45° on a row of fan-shaped holes. The results showed only small increase of the laterally-averaged film effectiveness for low blowing rates and a moderate compound angle of 30°. Further increase of the lateral hole inclination resulted into decreasing film effectiveness values. For larger blowing rates, the influence on the lateral effectiveness is negligibly small. In contrast, differences can be observed with increasing compound angle. The spatial film effectiveness distribution showed regions of locally enhanced film effectiveness that occur downstream of the rounded edge of the diffuser exit. With increased jet momentum, this enhancement is reduced again. Considering the heat transfer coefficient, significant heat transfer augmentation for injection at high blowing ratios and large compound angles of 45° was observed. This was due to increased turbulence induced by the coolant flow exiting the hole via the rounded diffuser corner.

2.3.2 Effect of Density Ratio

In order to achieve engine-like density ratios between the coolant and the mainstream, different experimental approaches can be considered. One is to perform measurements with large temperature differences between the main stream and the injected coolant flow. The coolant is cooled down to cryogenic temperatures using liquid nitrogen as e.g. done by Cutbird and Bogard [29], Sinha et al. [121] or Schmidt et al. [117]). In the case of cryogenically cooled air, the low temperature of the dense jets means a potential for H_2O and CO_2 to solidify and accumulate in the coolant supply and in the test section. Accumulated frost inside the holes and around the injection locations has to be removed in order to maintain consistent performance of the test rig. Sealing and drying techniques have to be applied to minimize air infiltration and to reduce frosting potential. Same temperature ratios can be achieved by heating the main stream to temperature levels of about 500°C, as done by Gritsch et al. [47], Brauckmann and Wolfersdorf [17] and Baldauf et al. [5]. A third approach uses a foreign gas with higher molecular weight, as e.g. Carbon dioxide (CO_2), Freon or Refrigerant-12 (R_{12}). Results on flat plate experiments using foreign gas injection for film-cooling can be found in Ekkad et al. [32], Goldstein et al. [42], Yuen et al. [143], Yuen [142], Pedersen et al. [102] and Teekaram et al. [125].

The density ratio affects indirectly the film-cooling injection. An increase in density ratio results in decreasing momentum flux ratios for a maintained blowing rate. As a result of that, variation in coolant-to-main flow density ratio can be used to reduce jet lift-off. This means also that the optimum blowing rate for attached cooling films will be moved to higher magnitudes with increasing density ratio. In this context, the work of Goldstein et al. [42] published in 1974 is here probably one of the earliest but most ground-breaking investigations on discrete film-cooling. The authors studied the effect of density ratio and mainstream boundary layer thickness on cylindrical and diffuser hole geometries. Results were discussed in terms of centerline and laterally averaged film effectiveness for a row of 35° inclined holes with three diameter spacing. Refrigerant-12 and air is used as coolant to produce a significant difference in coolant density.

The experiments indicate that the use of the gas R-12 gives significant increase in film-cooling effectiveness for cylindrical holes at high blowing rates compared to air injection. The effect is found to be smaller for low blowing rates as the jets stay closer to the wall rather than penetrating into the cross flow. Comparing centerline film effectiveness for air and Freon-gas injection at $x/D = 5.6$ and 6.7, it can be observed that film effectiveness (plotted as a function of momentum flux ratio) is significantly higher for the case of the high-density coolant. However, it should be noted that the horizontal position of the maximum shows good agreement for the low- and high-density case. Thus the data should be correlated better by taking into account the momentum instead of the mass velocity ratio as the key parameter for film injection when density ratio differences are considered.

Pedersen et al. [102] studied the effect of large density differences on a row of 35° inclined cylindrical holes by injecting Helium, R-12 and CO_2 into the mainstream. The heavier gases were mixed with air whether in small amounts to produce a constant property conditions or in larger concentrations to achieve differences in coolant density. The heat-transfer analogy approach was used to measure film effectiveness by determining local gas concentration through taps in the test plate. Centerline as well as lateral averaged film effectiveness is shown for an extensive range of density ratios D_R from 0.7 to 4 producing data sets to study the effects of velocity ratio, blowing ratio and momentum ratio on effectiveness. It is observed that the effect of density ratio is comparatively small for the studied low blowing rates of 0.21 and 0.52.

With increasing density ratio, its influence becomes more apparent and shifts the peak effectiveness toward higher blowing rates. Dense jets tends to stay closer to the wall with less trend for detaching and penetrating into the cross flow compared to the jets with low density at the same blowing rate. The highest film effectiveness values were obtained for the density ratio case of 4.17 and appeared to increase at very high blowing ratios. To quantify the effect of the density ratio, centerline film effectiveness data is extracted from this investigation for two specific density ratios of 1.06 and 1.57. The relative change equals about 48 % and is typical of what is achievable using Carbon dioxide for coolant injection. Considering the dimensionless streamwise distance of $x/D = 20$, the increase in film effectiveness is about 55 % for a blowing rate of 1.05 in the high-density ratio case. The benefit drops to less than 22 % with higher blowing rate of 1.96.

Also, Sinha et al. [121] studied film-cooling effectiveness with variable density ratio using a row of 7 inclined holes which inject cryogenically cooled air across an adiabatic test plate. Surface temperatures are measured using surface thermocouple arrangement along the centerline and in lateral direction. In general, the authors concluded that the effect of the density ratio on film effectiveness may not be scaled with mass flux ratio, velocity ratio or momentum flux ratio due to the detachment-reattachment nature of the jets with varying injection conditions. For low blowing rates where the jets remain attached to the wall, the density ratio has less effect on the film effectiveness which therefore scales with the mass flux ratio. At higher blowing rates, the tendency of the coolant jets to detach and reattach onto the surface becomes more important. Here the film effectiveness scales with the momentum flux ratio. For a change in density ratio from 1.2 to 1.6 (equals about 33 % relative increase in coolant density ratio), the film effectiveness increases about 60 % for $M = 1.0$ at streamwise distances of $x/D = 10$.

The investigations done by the group of Ekkad et al. [32] show film effectiveness and heat transfer coefficients for a row of 35° inclined holes with various compound angles. A transient liquid-crystal method was used to measure local heat transfer and film-cooling effectiveness. Air and CO_2 were used as coolant to produce differences in density ratio. High-dense coolant produced higher laterally averaged film effectiveness at high blowing rates compared to the case with air injection. The effect is less pronounced at lower blowing rates.

2.3.3 Effect of Free-Stream Turbulence Intensity

The effect of the main flow turbulence intensity has been investigated by a number of research groups (Kadotani and Goldstein [62,63], Jumper [60], Schmidt and Bogard [115], Bons et al. [13] and others) on cylindrical film holes. In general, research on the effects of the elevated free-stream turbulence in the range of 3 to 17 % for round holes has shown significant reduction of film effectiveness for low blowing rates where the film jets stay attached to the surface. The interaction and mixing process of the film jets in lateral and wall-normal direction is intensified with elevated free-stream turbulence. It weakens the film-cooling performance in streamwise direction and leads therefore to 20 - 40 % reduction in film effectiveness. At higher blowing ratios, round holes experience a slight increase in film effectiveness with higher turbulence levels. It is caused by the intensified transport of the coolant back toward the surface. In these cases, the jet strength and jet turbulence dominates the response to the elevated free-stream turbulence.

Bons et al. [13] studied the effect of high free-stream turbulence on adiabatic film effectiveness for a single row of 5 holes inclined 35° in streamwise direction. Data is presented for elevated free-stream turbulence ranging from 0.9 to 17 % at constant Reynolds number $Re_D = 19000$ based on the hole diameter. Centerline film-cooling effectiveness is presented for four different free-stream turbulence levels at a blowing rate of 0.75 and 1.5. Observing the results, distinct reduction of the centerline film effectiveness by up to 70 % is seen in the region immediately downstream of the hole injection. The effect of increased main flow turbulence steadily decreases in streamwise direction with lesser decrease in film effectiveness. At low blowing rates, the jet momentum is small keeping the films close to the surface. With weak resistance of the coolant against the enhanced mixing in the shear layers at high free-stream turbulence, the large fluctuations in the main flow field break down the jet structure quickly. At higher blowing rates, the jets appear to be stronger at the injection position due to the increased momentum. The effect of turbulence is small compared to the low blowing ratio case. The jets with high momentum sustain under the high free-stream turbulence and lift-off from the surface. Farther downstream, the jets reattach to the wall and dissipate faster under the elevated main flow turbulence. This leads to reduced film effectiveness downstream of $x/D = 40$.

Schmidt and Bogard [115] studied the effect of increased mainstream turbulence under realistic conditions of the density ratio. Film-cooling effectiveness and heat transfer coefficients were measured with high free-stream turbulence levels of 0.3, 10 and 17 % applying a coolant density ratio of $D_R = 2$. The integral turbulent length scale was three hole diameter for highly turbulent conditions. At a momentum flux ratio of 0.2 (equals $M = 0.6$ with $D_R = 2$), the film effectiveness is reduced by a factor of two close to injection location and essentially to $\eta = 0$ for $x/D > 20$. It was also concluded by the authors that the optimal momentum ratio for the high turbulence condition (10 %) is one order of magnitude larger ($I = 1.1$) than for the low turbulence case. The additional jet dispersion caused by the high free-stream turbulence leads to coolant that is transported back toward the surface raising the mixing temperature near the wall. No further differences in film effectiveness were observed when the turbulence intensity was increased from 10 to 17 %.

Saumweber et al. [112] studied the effect of three different free-stream turbulence levels of 3.6, 7.5 and 11 % on shaped holes. The authors presented film effectiveness and heat transfer measurements on a row of three cylindrical holes used as reference and two types of shaped hole geometries. The length of the metering section for both shaped holes is two diameters. The first one, denoted as fan-shaped hole, has an expanded exit portion in lateral direction with an opening angle of 14°. Additionally, the laid-back fan-shaped hole expands in axial direction over a 1D-portion with an opening angle of 15°. The laterally averaged film effectiveness data for the round hole shows differences with increased turbulence at low and high blowing ratios. The effect of high free-stream turbulence on shaped holes is detrimental. It shows less reduced laterally averaged film effectiveness up to 30 % when the turbulence is increased from 3.6 to 11 %. The effect is more pronounced for lower blowing rates. Furthermore, the authors state that the effect of integral length scale, as it was varied from $2.1D$ to $3.5D$, is less significant on cylindrical holes at constant turbulence intensity of 5 %.

Additionally, Burd et al. [23] and Kohli and Bogard [72] conducted flow field and temperature field measurements to quantify the effect of free-stream turbulence on film-cooling jets. Hot wire anemometer measurements were done by Burd et al. [23] in order to study the flow field downstream of a row of cylindrical film-cooling holes at two turbulence levels at 0.5 and 12 %. The authors stated significant differences in the flow field downstream of the holes for cases with varied length-to-diameter ratio from 2.3 to 7. The pronounced differences between short hole and long hole injection diminish as high turbulence conditions are present in the main stream. It was concluded by the authors that the L/D-ratio has to be taken into account when turbulence effects shall be considered. Kohli and Bogard [72] used a high frequency response temperature probe to investigate the dispersion of the coolant jet injection from cylindrical holes for two different turbulence levels at 0.5 and 20 %.

2.3.4 Effect of Surface Roughness

Most investigations on film-cooling performance deal with smooth surfaces which would be representative of new turbine surfaces. During operation of the gas turbine, the turbine components normally become rough due to deposition and erosion. Increased surface roughness may lead to earlier boundary layer transition, thickening of the boundary layer and increased turbulent mixing in the boundary layer. Investigations on the effect of surface roughness are conducted by Goldstein et al. [41], Schmidt and Bogard [115], Schmidt et al. [116] and Weigand et al. [135]. Diverse methods are used to simulate surface roughness where qualitative and quantitative differences can be observed in the experimental data. Schmidt et al. [116] used an array of conical elements with a maximum roughness height of $0.4D$. It corresponds to an equivalent sand grain roughness of $R_{sand} \approx 100$. It was found that the roughness downstream of cooling holes has only small effect on the laterally averaged film effectiveness. It leads to less than 10 % decrease for low momentum ratios and less than 5 % increase for high momentum ratios. Higher surface roughness caused a significant enhancement in heat transfer of the factor two due to the enlarged test surface. The ratio of the heat transfer coefficient with and without coolant injection shows only negligible change.

2.3.5 Effect of Boundary Layer Thickness

The effect of the boundary layer thickness on local film effectiveness was investigated by Kadotani and Goldstein [61]. It occurred only in the region immediately downstream of the holes where increasing displacement thickness ($\delta^*/D = 0.5$ ... $\delta^*/D = 0.2$) caused slightly enhanced film-cooling effectiveness between the holes and reduced centerline effectiveness in streamwise direction. Due to the countering effects, the laterally averaged film effectiveness level remained unchanged. The work of Liess [79] indicated that for ratios of the displacement thickness-to-hole diameter of less than 0.2, the film effectiveness level remains unchanged. The experimental data shows reduced laterally film effectiveness for increasing displacement thickness of the free-stream boundary layer.

2.3.6 Effect of Mach Number

Many previous investigations that address the effect of Mach number are designed as flat plate experiments designed and operate at low or high speed flow conditions. There are only a limited number of film-cooling experiments where results were obtained at supersonic conditions. Gritsch et al. [48] studied the effect of an external Mach number variation of 0.3, 0.6 and 1.2 for three different hole geometries. The authors determined adiabatic film effectiveness for film-cooling injection through a single cylindrical hole, lateral-expanded diffuser hole and a laid-back fan-shaped hole. Their results of the laterally averaged film effectiveness showed hardly any effect of the crossflow Mach number at all blowing rates for the subsonic cases of $Ma = 0.3$ and 0.6. Increased laterally averaged film effectiveness was found for $Ma = 1.2$ which was attributed to the shock structure generated by injecting coolant into a supersonic flow field. They stated that the generation of a bow shock and the resulting shock-induced pressure field can have the tendency to turn the jet toward the surface. These findings coincide with the results of Liess [79] who performed film effectiveness measurements on a row of coolant holes for Mach numbers of 0.3, 0.6 and 0.9. The experiments showed that the variation in Mach number had no effect on film effectiveness.

2.4 Objective of the Present Work

When applying film-cooling, the main target is to minimize the coolant consumption of the turbine. This is achieved by optimal distribution of the cooling fluid in the different cooled parts of the vane or blade. Although the basic principles of film-cooling seem to be quite simple, the resulting three-dimensional flow field and the complex thermal conditions that are created during to the interaction of two fluids at different velocities and temperatures are difficult to describe analytically. Primarily, hub and casing platforms occupy the largest area portion of all cooled wall surfaces in a first high-pressure turbine vane. Therefore, a large number of holes and subsequently a large amount of coolant is required. This leads to great challenges in achieving an optimum cooling design in terms of cooling performance, coolant amount, structural integrity and manufacturing costs.

On the platforms, the coolant injection takes places into the highly distorted, three-dimensional wall boundary layer where the near-wall region is dominated by intense cross flow gradients between the adjacent airfoils. At the injection locations, interaction and mixing processes in the shear layers between the coolant and the hot gas cause rapid consumption of the cooling fluid during its streamwise convection. This leads to a steep decrease of the jet's cooling performance. The mixing intensity depends on the specified hole geometry and the blowing parameters as well as on the hot gas flow field. Since the individual film hole rows of a platform configuration are placed in different axial and lateral positions, the acceleration of the hot gas and the changed wall pressure field has significant influence on the film jet injection as well as on the row-by-row coolant distribution. Surface curvature but also rotation of the components additionally alters the mixing process.

In turbomachines, endwall film-cooling is associated with even more complex interaction and mixing due to the near-wall flow field that is dominated by secondary flow structures and is therefore of three-dimensional, vortical and unsteady character. The radial and circumferential pressure gradient is the driving potential for this phenomenon. The vortical movement and the strength of the secondary flow is intensified with increased airfoil loading. Indeed, previous studies have shown that the three-dimensional near-wall flow field can have tremendous influence on coolant injection. But essentially no results were found for increased airfoil loading or off-design conditions. Pre-mixed combustion produce strong swirling flows to attain ultra low nitrogen oxide emission. Using strongly curved transition pieces to connect the combustor and the turbine inlet at different radii may induce additional vortical flow pattern inside those channels. Placing the first vane in such an environment, the inlet flow angle and hence the profile loading may be changed significantly. During engine operation of the first turbine vane, variation of the inlet flow angle of about 10° to 15° can occur. This leads to increased vane turning and significant enhancement of the secondary flow.

Therefore, the primary aim of the present work is to investigate platform cooling with variation of cascade loading. The experiments are performed in a straight cascade with a fully-cooled platform configuration. The hole scheme incorporates a large number of film rows distributed all over the passage. The investigated profile and the platform cooling geometry is based on a near-hub section of the nozzle guide vane of a highly-loaded gas turbine. The experiments are carried for different loading cases by changing the incidence between i = -5° (low loading) to +40° (maximum loading). The surface temperature distribution with film-cooling injection is determined using temperature-sensitive paints (TSP). This measurement technique is tested and validated on a basic experimental setup using flat plate film-cooling. Due to the large temperature difference between the hot gas and the coolant in real engines of about 0.4, the density ratio is approximately 2.0 or higher. To simulate this effect, carbon dioxide was used as coolant with that elevated density ratios of about 1.4 can be established.

Although a low thermal conductivity material is used for the endwall test plate, the measured temperature fields showed influences of three-dimensional heat conduction inside the test plate. To measure film effectiveness and the heat transfer separately, an adiabatic test surface is needed. Therefore the effects of heat conduction are modeled by a thermal analysis using a Finite-Element method. A methodology including comprehensive data reduction is developed to correct the wall temperature measurements toward adiabatic conditions. This approach uses the calculated convective heat flux pattern which is superimposed onto the measured wall temperatures. The adiabatic wall temperature is then evaluated by a one-dimensional heat balance. In order to finally obtain the adiabatic film-cooling effectiveness, the heat loss inside the holes and subsequently the film-discharge temperature at the hole exit is estimated.

3 Experimental Setup

3.1 Test Rig for Flat Plate Film-Cooling

Basic flat plate film-cooling experiments were conducted for validation purposes of the temperature-sensitive paint technique in a subsonic open-circuit wind tunnel. Fig. 3.1 shows the schematic test rig setup. Due to the basic character of this conventional experimental setup, film-cooling parameters can be precisely controlled and its effect on the film-cooling process can be investigated in detail.

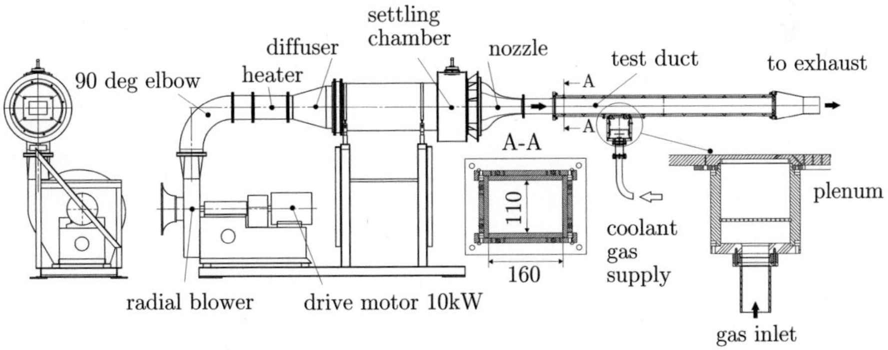

Figure 3.1: Schematic of flat plate test rig

The wind tunnel with the film-cooling setup was formerly used for probe calibration purposes. Here, ambient air is compressed by a blower with an electrical power input of $10\,kW$. The air is then routed through supply pipes toward an electrical pipe heater which is used to maintain the main flow temperature level at varying ambient conditions. The main flow temperature is held constant at $T_m = 30\,°C$ for all investigated cases. Downstream of the heater, the flow passes an assembly of screens and honeycombs which are used to homogenize the flow before it enters the settling chamber. From there, the flow is accelerated by a nozzle with a 7:1 contraction ratio. Before entering the rectangular test section measuring $0.16\,m$ x $0.11\,m$ in cross-section, the main flow total and static pressure p_t, p_m as well as the total temperature $T_{t,m}$ are determined by a Prandtl-probe and a calibrated thermocouple. The static wall pressure is measured by a pressure tap. The mainstream velocity is maintained at 22m/s at test section entry which equals a Reynolds number $Re_{m,D} = 4620$ based on the hole diameter D.

The test-plate surface is instrumented with a row of 10 thermocouples down its centerline. Furthermore, several thermocouples are embedded into the test surface off centerline to check for symmetry. The static wall pressure of the main flow is measured by pressure taps placed at a distance of x/D = -17 upstream of the holes. The injection plate can be exchanged with other test plates containing different film hole geometries. All parts of the test section are machined from Perspex material with a low thermal conductivity of $\lambda = 0.19\,W/(mK)$ to minimize thermal losses and heat conduction through the test plate. The test-duct's upper wall is equipped with two measurement locations upstream of the holes at $x_1 = -0.355\,m$ and $x_2 = -0.180\,m$ with respect to the hole center. It equals a normalized distance from the holes of x/D = -118 and -60 with regard to the hole diameter of the cylindrical holes. Either temperature probes, pneumatic probes or a hot wire probe can then be inserted through the upper plate to obtain main flow velocity profiles, temperature or turbulence intensity distributions. The test section is built onto a rack along with two linear traversing units. The CCD camera is positioned above the test section with a repeatability of $\pm\,0.02\,mm$.

3.1.1 Film Hole Geometry

Two different hole geometries were selected for the flat plate experiments. The film-cooling experiments are carried out for plain cylindrical and diffuser holes. Table 3.1 summarizes the geometrical parameters of the used film holes along with the fluid boundary conditions of the main flow. The film holes are machined into the perspex test plate. The diffuser exits are sharp-edged. Flat plate experiments were performed on a film row with 7 cylindrical holes. The hole geometry is based on the dimensionless parameters given in Ekkad et al. [31]. The cylindrical holes are inclined at 35° with regard to the streamwise flow direction. The dimensionless length-to-diameter ratio is set to $L_h/D = 4.6$ and the pitch-to-diameter ratio is $P_h/D = 4$. The hole diameter was selected to D = 3mm.

The shaped holes are designed in accordance with the fan-shaped geometry investigated by Gritsch et al. [48] and Saumweber et al. [112]. It can be denoted as Type B, following the classification introduced by Bunker [21], due to its 2-stage diffuser. The first stage of the diffuser expands the cylindrical hole section first in lateral direction and is followed by an axial expansion toward the hole exit. The row consists of five shaped film holes with a hole diameter of D = 5mm. The inclination angle with regard to the test surface is 30°. The dimensionless total length-to-diameter ratio of the hole is $L_h/D = 6$ which results in an pitch-to-diameter ratio of $P_h/D = 4$. The length of the diffuser section including both lateral and axial expansion is $L_{fan}/D = 4$. The lateral diffuser works over a hole length of $3D$ and the axial diffuser part is $1D$ long. The diffuser features quite large opening angles of $\gamma_{lat} = 14°$ in lateral direction and $\gamma_{ax} = 15°$ in streamwise direction. The metering section measures a length-to-diameter ratio of $L_{met}/D = 2$. That would still be in agreement with the suggestion that the minimum length of the metering section should be at least 2D as given by Hay and Lampard [54]. It will

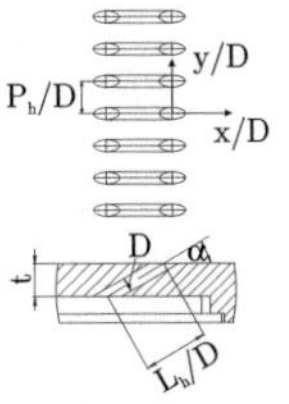

L_h/D	$L_{fan,ax}/D$	$L_{fan,lat}/D$	D	P_h/D	α	γ_{ax}	γ_{lat}	N
[-]	[-]	[-]	[mm]	[-]	[°]	[°]	[°]	[-]
4.6	-	-	3	4	30	-	-	7

u_m	δ^*/D	$Re_{m,D}$	Tu	T_R	D_R	M	I	gas
[m/s]	[-]	[-]	[%]	[-]	[-]	[-]	[-]	[-]
22	0.56	4620	1.5	1.11	0.90-1.37	0.35-2.0		air CO_2

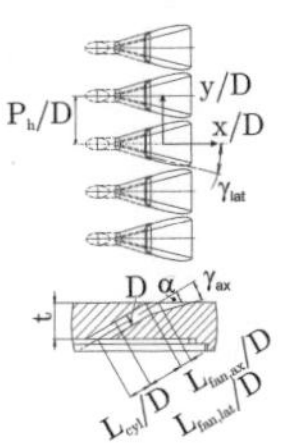

L_h/D	$L_{fan,ax}/D$	$L_{fan,lat}/D$	D	P_h/D	α	γ_{ax}	γ_{lat}	N
[-]	[-]	[-]	[mm]	[-]	[°]	[°]	[°]	[-]
6	1	3	5	4	30	15	14	5

u_m	δ^*/D	$Re_{m,D}$	Tu	T_R	D_R	M	I	gas
[m/s]	[-]	[-]	[%]	[-]	[-]	[-]	[-]	[-]
22	0.33	7250	1.5	1.11	0.90-1.37	0.35-2.0		air CO_2

Table 3.1: Summary of geometrical parameters for the cylindrical and diffuser holes used during the flat-plate film-cooling experiments

allow the occurring flow separation at the hole inlet to reattach to the wall before entering the diffuser. Thus, the operation of the diffuser is undisturbed by the inlet separation region which increases the efficiency of the following diffusion process.

The test plate is sprayed with 6-7 thin layers of UNI-TSP 01 which is commercially available from Innovative Scientific Solutions Incorporated. The paint is applied within a surface region of 180 mm x 110 mm downstream and 50 mm x 110 mm upstream of the holes. Due to the viewing field of the camera, the maximum measurement region measures $0 < x/D < 26$ in axial direction and $-10 < y/D < +10$ in pitchwise direction concentrating on the near-hole injection-region with the jet-main flow interaction. The view field includes the three center holes of the row. The thermocouples along the hole centerline are equally spaced at 8 hole diameters. Thermocouples embedded off the centerline position are arranged parallel to the hole center in pitchwise direction. The CCD-camera is mounted above the test duct onto the two linear traversing axes. The camera is set with a NIKON^{TM} Nikor $f = 58\,mm$ lens and focused on the test surface. The TSP-coated surface is constantly illuminated by the LED-light sources and the measurement images are acquired.

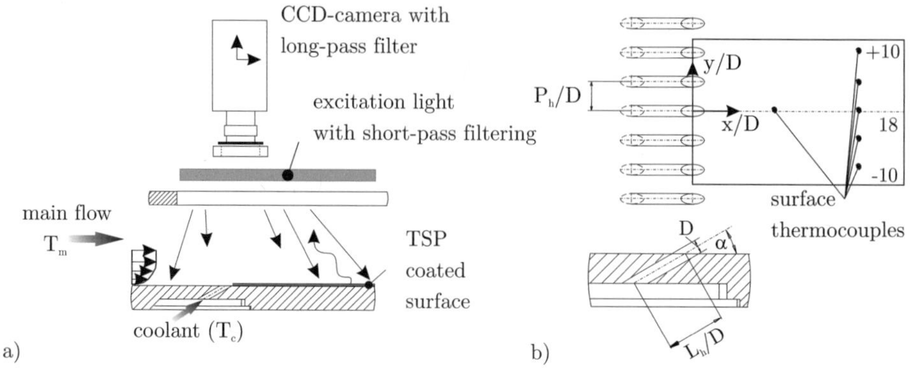

Figure 3.2: Setup of CCD-camera (a) and surface measurement region with thermocouple locations (b)

3.1.2 Coolant Supply

The coolant supply of the flat plate test rig is designed to represent a single supply line of the gas feed system for the endwall film-cooling investigations in the turbine cascade (see also section 3.2.3). It can either be operated with air supplied by a pressurized air system or with a foreign gas like CO_2 provided by gas bottles.The high bottle pressure of 100 *bar* is reduced by a pressure valve to operating pressures of 20 *bar* before the fluid enters the supply lines. For detection of CO_2 leakage in the laboratory, the installed alert system composes of several IR-sensors. It can automatically shut down the coolant flow by closing the solenoid valve at the gas bottle station. When the gas enters the laboratory, it passes a second pressure valve where the operating line pressure can be further reduced between 0 and 10 bar.

The coolant-supply system consists of a mass flow controller with a solenoid valve, a pipe heater and a plenum chamber. Before entering the plenum chamber, the flow rate is set by the mass flow controller from Alicat Inc. within the flow rate range of 0 ... 250 standard liter per minute (SLPM). The measurement uncertainty of the mass flow reading is ±0.5 % of the full scale range. The flow measuring device composes of a proportional valve and a laminar flow element (LFE). The proportional valve is electronically coupled with the LFE to adjust the actual flow rate to an user-defined value. The operating principle of the LFE is based on the physics of the Hagen-Poiseuille equation. It quantifies the linear relationship between pressure drop, flow rate and kinematic viscosity of a flow of laminar state inside a pipe. The laminar flow element forces the gas to a state of laminar flow beneath the Reynolds number threshold of 2000. The pressure drop over the LFE is measured within the laminar region. Finally, the viscosity is determined by measuring the temperature of the gas. For further details, the measurement principle of the LFE is described in chapter 4.2. The solenoid valve downstream

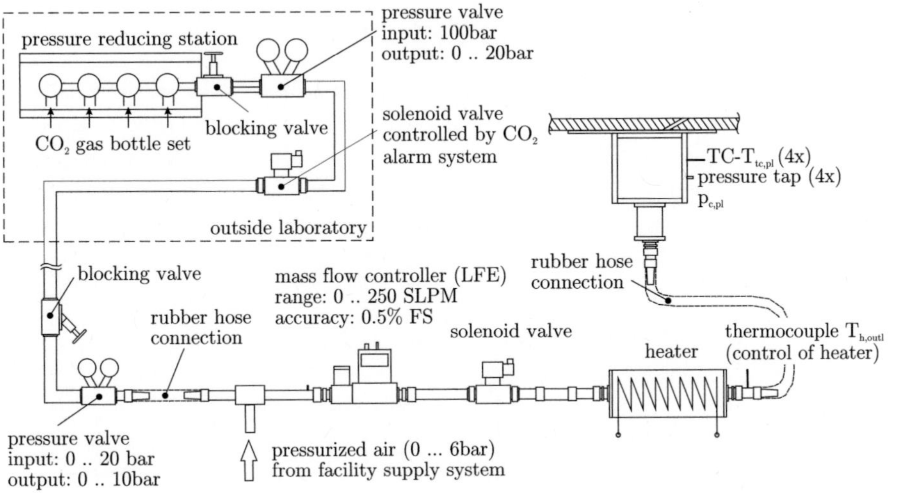

Figure 3.3: Schematic of coolant supply system

of the mass-flow meter is used to open or shut down the gas feeding to the plenum. The flow is then heated by an electrical heater which is automatically controlled by thyristor regulators. A thermocouple located at the heater exit is used to control the coolant temperature. The coolant temperature in the plenum is maintained at 65°C $\pm 0.4\,K$ which provides a sufficient temperature difference that is required for the TSP-measurements. At the plenum inlet, coolant is piped through a combination of a screen and a perforated plate to ensure a homogenized inflow. Thermocouples and pressure taps inside the plenum chamber allows the measurement of the total plenum pressure $p_{t,pl}$ and the coolant total temperature $T_{t,pl}$. Assuming low flow velocities in the plenum, the total pressure and total temperature can be replaced by their static values. The coolant is finally injected through the row of film-cooling holes into the main flow at a normalized distance of x/D = -177 downstream of the test-section entrance.

3.2 Test Rig for Endwall Film-Cooling

The endwall film-cooling experiments are carried out in the Low Speed Research Cascade Wind Tunnel (LSRCW) of the Technische Universität Dresden. The wind tunnel is designed as an open-cycle test rig and can deliver a maximum exit velocity of $70\,m/s$ which is translated into a maximum Mach number of $Ma_2 = 0.2$ at the cascade exit. It results in an exit Reynolds number based on the chord length of $8.1{\cdot}10^5$. The inlet Mach number for the operating point of the profile is 0.05 which equals an inlet Reynolds number based on the chord length of $Re_1 = 2.6{\cdot}10^5$.

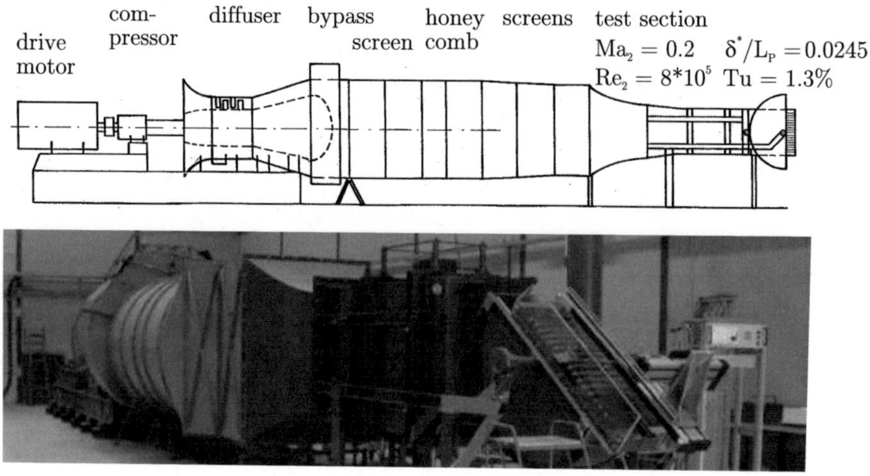

Figure 3.4: Schematic of Low Speed Research Cascade Wind Tunnel

Air is compressed by a 2-stage compressor driven by a motor with a maximum power input of $150\,kW$. It passes a system of flow-conditioning devices including a diffuser, a bypass, an assembly of a honeycomb and several screens to reduce the turbulence level to about 1.3 % within the core of the inlet flow. The bypass features a circumferential gap and two adjustable hatches which are used to adjust the leakage flow through the bypass gap. Therefore the pressure inside the wind tunnel and the loading of the compressor can be controlled. Downstream of the bypass, the compressed fluid is straightened by a large screen with a cross-sectional area of $3.05\,m^2$ and a honeycomb located downstream of that screen. The circular cross-section with a diameter of $1.97\,m$ is then transferred to a rectangular cross section of $1.93\,m$ x $1.41\,m$. Two additional screens with an area of $2.70m^2$ each are installed downstream to further homogenize the flow.

Before entering the test section, the main flow is accelerated by a contraction nozzle into the $0.3\,m$ x $0.94\,m$ turnable test section. This allows adjusting of the inlet flow angle to the actual cascade configuration between a range of 119° to 30°. This equals a cascade incidence from -19° to +60° which provides significantly higher blade loading. The contraction of the flow is carried out in two steps. First the upper and lower walls are lowered to a variable test section height ranging from 0.25 m to 0.94 m. Then the sidewalls are transferred to the test section width measuring 0.3 m. This results into a maximum contraction ratio of 9:1. After passing the cascade setup, the air expands to ambient conditions. A Prandtl-probe together with a static pressure probe and a thermocouple are installed at x/L_{ax} = -9 upstream of the cascade inlet. The measured velocity head $p_{t0} - p_0$ obtained by the Prandtl-probe in combination with the static pressure p_0 and the inlet temperature T_{t0} is used to determine the cascade inlet flow velocity.

3.2.1 Airfoil Definition

The cascade is built of 6 airfoils with an aspect ratio of 1.58 and a pitch-to-chord length ratio of 0.968. The large aspect ratio was selected to ensure a 2D-passage flow basically over a large spanwise portion of around 40 % blade height. The airfoil geometry is based on a nozzle guide vane of a highly-loaded stationary gas turbine. The profile at midspan position represents a two-dimensional section of the vane contour near the platform. The profile is extended as straight cylindrical section over 88 % of the airfoil span.

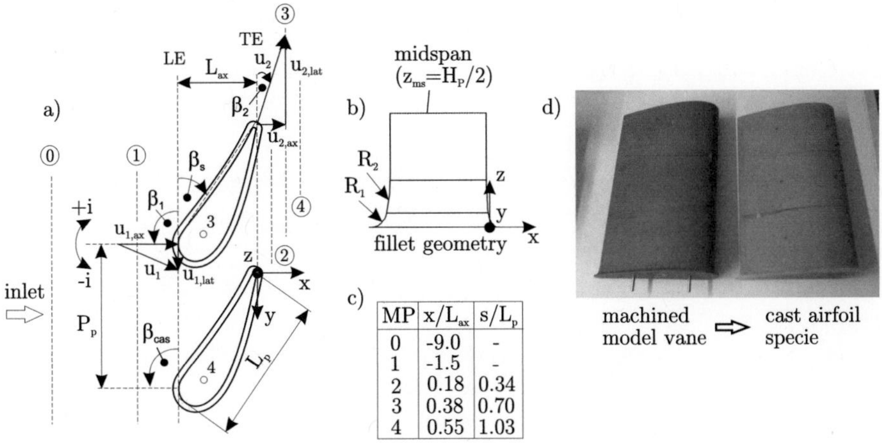

MP	x/L_{ax}	s/L_p
0	-9.0	-
1	-1.5	-
2	0.18	0.34
3	0.38	0.70
4	0.55	1.03

Figure 3.5: Definition of airfoil geometry for the investigated turbine vane

Near the platforms the vane sleeve is transferred onto the platform by a fillet geometry. A blunt trailing edge of 2.6 % of the profile length is incorporated into the airfoil that contains pin fin and trailing edge cooling features in the real engine vane. The fillet starts tangentially at the vane hub and is merged into the cylindrical profile by two radii ratios $R_1/L_p = 0.033$ and

$R_2/L_p = 0.33$. The transition is completed at a distance from the endwall of 6 % blade span. For the film-cooling investigations, the vane root profile extends spanwise for 1.5 mm which fits into a recessed pocket in the endwall test plate. This geometrical feature fixes the airfoil stagger and its position relative to the film-hole locations which allows similar and reproducible vane setting during the cascade assembly.

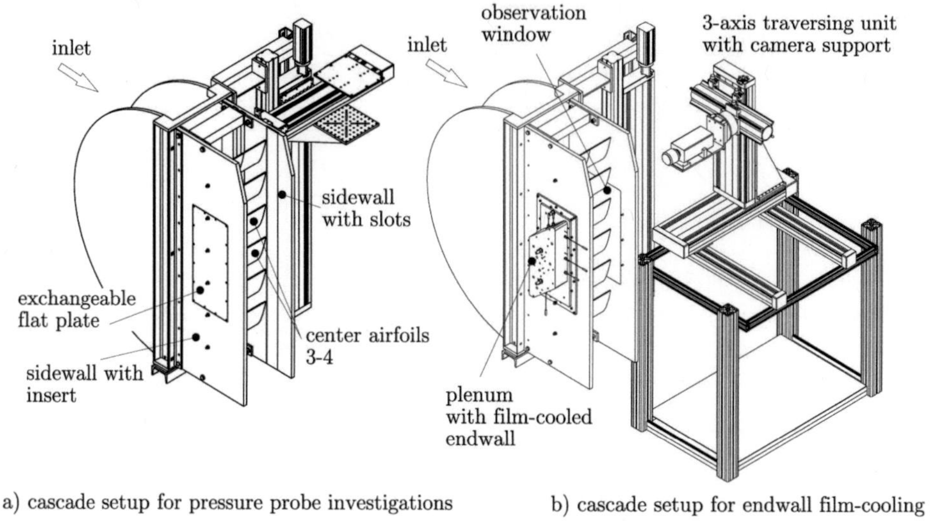

Figure 3.6: Cascade configuration for a)aerodynamic and b) film effectiveness measurements on the endwall

The cascade flow field is obtained upstream and downstream of the cascade. The location of the measurement planes can be depicted in Fig. 3.5. The inlet boundary-layer thickness and the turbulence intensity were measured at a location x/L_{ax} = -1.52 upstream of the trailing edge using a pitot and a single hot-wire probe respectively. The total pressure losses and the corresponding flow field at the cascade exit are obtained for three different axial planes at x/L_{ax} = 0.18 - 0.55 with regard to the trailing edge coordinate system. It can be translated into normalized streamwise distances of s/L_p = 0.39, 0.70 and 1.0 with regard to the vane chord. These defined planes are maintained for all test cases if not mentioned otherwise.

The airfoils are manufactured by casting. Prior to the casting process, a model including the fillet was manufactured using high-precision 3 axis-CNC machining with tolerances less than $\pm 0.05\,mm$. Then a silicon mold was cast around the model blade which provides the negative vane contour for the later casting process. A stiffening inlay core was added to the mold to achieve uniform surface shrinking during cool-down and enhanced stability during wind tunnel operation. To reduce thermal losses, the vanes are cast using a transparent epoxy resin with low thermal conductivity. The deviation of the cast airfoil surface contour with regard to the electronic vane coordinates is evaluated by an in-house 3D-Laser scanning system. The averaged deviation is found to be 0.12 mm. The maximum is slightly less than 0.25 mm and is located on the suction side shoulder of the airfoil at 40 % chord.

The sidewalls were machined from perspex to minimize heat conduction effects in the test rig components and to provide optical access during the experiments. Two configurations of the cascade setup have been selected to achieve maximum variability and minimum modifications between the measurement campaigns. The primary hardware of the cascade with its two configurations is shown in Fig. 3.6. For the aerodynamic investigations, the right sidewall is equipped with machined slots of different sizes for insertion of the pneumatic probes into the main flow field. The slots are sealed with lineal-type seals to minimize air leakage. The opposite sidewall contains a machined breakout to mount different types of flat test plates. Plates containing wall pressure taps or the endwall film hole configuration can be mounted without disassembling the entire cascade setup. Identical dimensions of the test plates allows rapid and effortless conversion.

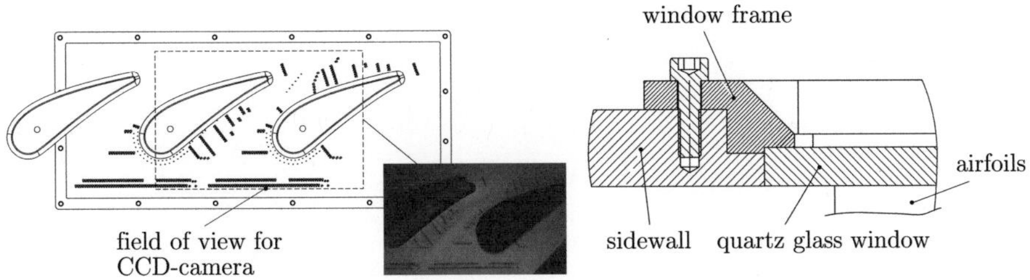

Figure 3.7: Window access for TSP-camera and optical measurement techniques

For the endwall film-cooling experiments, the right sidewall with the slots is replaced by a dimensionally similar wall containing a window recess. The optical access to the test section is provided by a 580 mm x 250 mm quartz-glass window showing the complete center section of the cascade with the adjacent center airfoils and the filmed endwall (as shown in Fig. 3.7). The window was made from SILUX4 by sandjet-cutting at SICO Ltd., Austria. It has an optical transmission range from 290 nm to 2800 nm and a very low content of bubbles or inclusions. The window is glued to a supporting PVC-plastics frame. The left sidewall contains the endwall test plate with the film-cooling holes which is mounted onto the casing of the coolant plenum and onto the sidewall. All components which are illuminated by the TSP-lighting system are painted black to receive higher contrast and reduce reflections. The cascade sidewalls are extended 0.20 m from the cascade exit to ensure guidance to the exiting flow, hence a more periodic flow field.

3.2.2 Film Hole Configuration on Cascade Endwall

The investigated film hole configuration is based on the platform cooling scheme of a highly-loaded gas turbine first vane. The scheme includes shaped film holes, denoted as Type A according to the classification proposed by Bunker [21]. The fan geometry contains a combined diffuser section with lateral and axial expansion. The holes are incorporated in an exchangeable Perspex test plate with a low thermal conductivity of $\lambda = 0.19\, W/(mK)$. The material is chosen to reduce the thermal losses and to keep the heat flow through the test plate as low as possible.

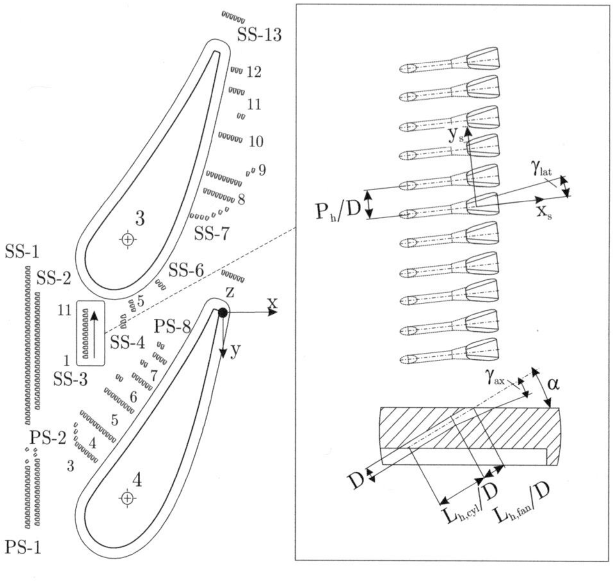

Figure 3.8: Endwall cooling scheme with hole geometry for row SS-3

For sufficient platform cooling, the film-hole rows are distributed across the entire passage. Individual film holes are combined to different row groups. One group is placed along the suction side far upstream as well as downstream of the airfoil throat denoted as SS-3 to SS-13. Furthermore several rows are placed upstream of the leading edge to provide sufficient cooling to the inlet endwall region (SS-1 to PS-2). These holes are expected to be strongly influenced by the developing secondary flow structures. Five rows are positioned on the endwall region along the pressure side (PS-4 to PS-8) which would otherwise be left uncovered of coolant. This is due to the strong near-wall crossflow which transports the cooling fluid toward SS of the adjacent vane. Each group of film rows is supplied by a separate plenum chamber (shown in Fig. 3.9a). It is equipped with a thermocouple and pressure taps to control the setting of the coolant mass flow rates supplied to each row. Each plenum chamber is connected to the gas feed system by a individual supply line. It allows the setting of different coolant mass flow rates per plenum chamber for either single or multiple row injection. For coolant injection through a single row, all other rows of the group are closed with strongly adhesive tape.

a) coolant plenum with different supply cavities

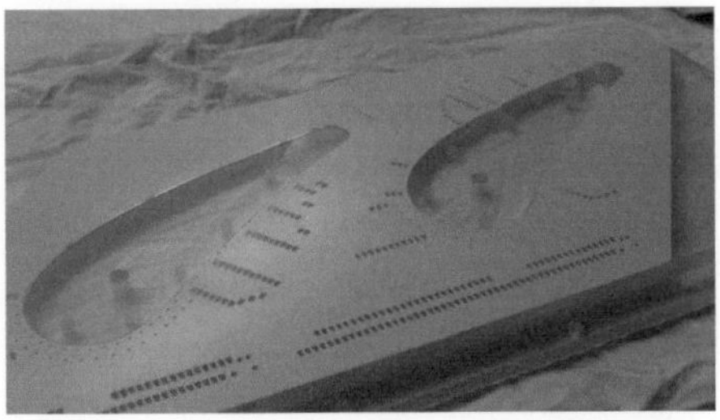

b) film-cooled endwall with applied TSP-paint

Figure 3.9: Coolant supply components machined from Perspex: plenum (a) and endwall test plate with film-cooling holes (b)

The TSP-paint is applied onto the endwall surface in several layers dispensed from aerosol cans with 6-8 layers until satisfactorily coverage is reached (the finished endwall surface with TSP can be seen in Fig. 3.9b). Prior to the spraying process, the endwall test-plate is carefully prepared by cleaning the plate surface. Furthermore the fillet recesses and the diffuser exits were closed with plasticine to avoid paint deposition inside those shallow trenches. The holes were then re-opened after drying of about two hours to finish the paint application.

A definition of the geometrical parameters for the used shaped holes should be given here, taking row SS-3 as representative example. This row is located near the suction side of the vane and consists of 11 fan-shaped holes with a total length-to-diameter ratio L_h/D of 16. The diffuser is designed with a moderate opening angle in lateral and streamwise direction. The holes are inclined at slightly less than 30° with regard to the streamwise direction. The pitch-to-diameter ratio P_h/D is 4. The compound angle of the holes is less than 6° with regard to the axial direction. The dimensionless length-to-diameter ratio of the metering section is about 3/4th of the total hole length ratio whereas the diffusion process takes place over the last 3 hole diameters.

3.2.3 Coolant Supply System

The adequate control of film-injection parameters and temperature is crucial for the successful performance of the film-cooling experiment under such highly complex flow conditions as on the endwall of a turbine cascade rig. The gas delivery system (see Fig. 3.10) is designed to operate with different types of gas or with a mixture of two gases at variable mass fraction. Either air or a foreign gas (in this case: CO_2) can be used for coolant injection. A mixing station is

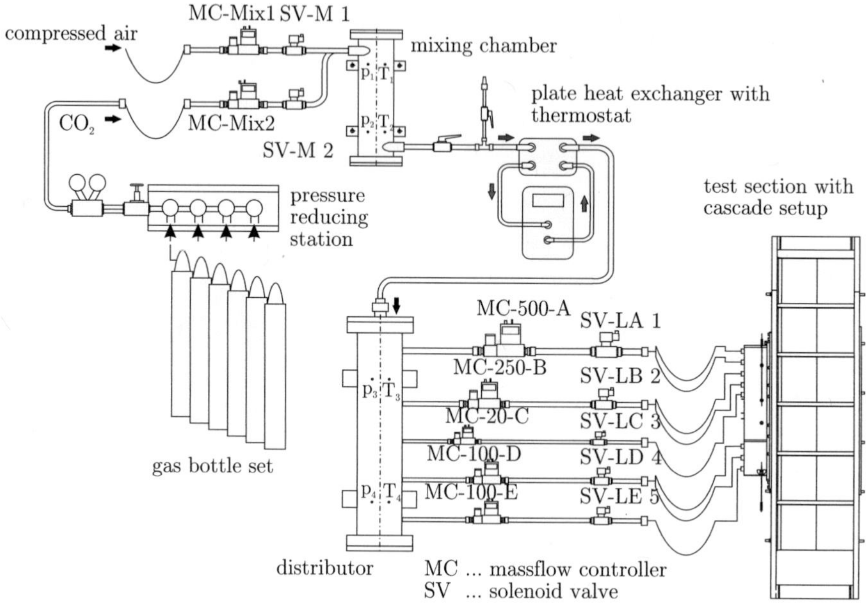

Figure 3.10: Schematic of coolant supply for endwall film-cooling investigations

installed upstream of the heat-exchanger system with flow-controller units at each gas inlet to provide mixing of the two gas paths. Compressed air is provided by a regulated Atlas Copco 180 kW screw-type compressor. Then the air is routed through a condense separator to reduce gas humidity and is directed into a large vessel with a volume capacity of $6m^3$. From the reservoir, the air passes a piping system with blocking valves and several locations to measure temperature and pressure before it is supplied to the mixing station. The foreign gas is provided by one or more gas bottle sets.

The containers are connected to a gas station which reduces the supply pressure down to 5 bar. For single gas injection, the mixing chamber is bypassed by routing the fluid through a heat exchanger into the distributor vessel. By opening the solenoid valve of a specific supply line and setting a required mass flow rate at the controller, the coolant flows into the corresponding plenum cavity and is injected through the selected film row. If a gas mixture should be set to simulate a certain density ratio for a film row, the two gas flows into the mixing chamber must be controlled. Therefore the flow controller in the supply line is replaced by a straight pipe connection. Otherwise the gas flow into the plenum is blocked by the proportional valve of the LFE. The total mass flow rate and the coolant density is then set and controlled by the mass fractions of both single gases at the mixing-chamber inlet. The resulting coolant density ratio of the mixed gas flow can be adjusted by adding CO_2 from 0 % to 100 %. Furthermore, the coolant properties at the hole exit are taken into account to calculate the mixed density ratio. They are determined from the measured plenum conditions assuming an isentropic, one-dimensional expansion across the hole.

After mixing of the two gases at isothermal conditions, the mixture is routed through a copper-molded plate heat-exchanger which is connected to a Lauda Integral Type T4600W thermostat. The thermostat can be operated at temperature levels between -25°C to 120°C and has a temperature stability of $\pm 0.2\,K$. An external Pt-100 resistance thermometer senses the outlet temperature of the gas exiting the heat exchanger.

line	mass flow controller	flow rate range		accuracy		
		[SLPM]	[g/s (CO_2)]	[%]	[SLPM]	[g/s (CO_2)]
A	MC-500	0 ... 500	0 ... 15	± 0.6	± 3	± 0.08
B	MC-250	0 ... 250	0 ... 7.5	± 0.6	± 1.5	± 0.045
C	MC-20	0 ... 20	0 ... 0.6	± 0.6	± 0.12	± 0.0036
D	MC-100	0 ... 100	0 ... 3	± 0.6	± 0.6	± 0.018
E	MC-100	0 ... 100	0 ... 3	± 0.6	± 0.6	± 0.018

Table 3.2: Mass flow controller units per supply line of the turbine coolant supply

The plenum and the endwall test plate are sealed using a silicon flat gasket of $3\,mm$ thickness. Each plenum chamber is equipped with a calibrated thermocouple and a pressure tap. Furthermore, temperature and pressure measurement locations are distributed in the mixing chamber and the distributor vessel. Each plenum chamber is connected to a supply line which contains a solenoid valve and a mass flow controller (Alicat Scientific Ltd.). These controllers consist of a proportional valve and a laminar flow element (see also chapter 4.2 for detailed description). All flow controllers are pre-calibrated and the accuracy of flow rate reading is given with 0.6 % full-scale of the measurement range for each unit. Repeatability of the flow rate setting is $\pm 0.2\,\%$. Units with different flow rate ranges are installed in the supply lines to account for various coolant flow consumption per plenum chamber (see Table 3.2).

3.2.4 Data Acquisition System

For this experimental setup, a multitude of measurement values have to be acquired. To control and record such large amount of data, a computer-controlled and automatic data acquisition system is needed. The data acquisition system of the LSCRW for the endwall cooling experiments is primarily composed of three parts: 1) the conventional data acquisition with a pressure and temperature measurement system including 4 high precision digital multimeter Keithley DMM 2700 with two multiplexer inserts and three pressure modules, 2) a rotary pressure system for measuring wall pressure fields and 3) the TSP measurement system with

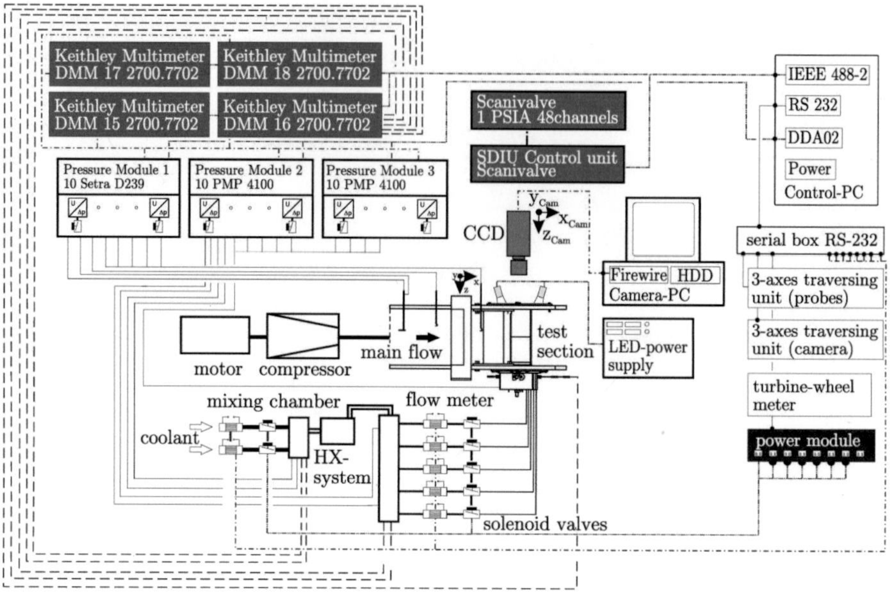

Figure 3.11: Schematic of data acquisition system at the cascade wind tunnel

CCD-camera and glass fiber-optical linked data storage. Each pressure module contains up to 10 pressure sensors. The rotary pressure system (Scanivalve SSS-48C) contains a differential pressure sensor with a pressure range of 1 PSID and a measurement accuracy of 0.06 % of the full measurement range. It is controlled via a SDIU control unit. The solenoid valves and the mass flow controller are remotely controlled via serial connections using a remote power control.

A fully automatic probe positioning system is placed at the wind tunnel which consists of two linear axis and a rotary unit. The linear traversing units are used for probe positioning in spanwise and pitchwise direction. The rotary motor allows rotation of the mounted pneumatic probe. Adjustment of the probe position in axial direction for measuring different inlet and outlet planes is performed manually. All units are operated by drive modules with a repeatability of $\pm 0.02\,mm$ and $\pm 0.03°$, respectively. The scientific CCD-camera of the TSP-measurement system is set onto a separate linear-traversing system with three axes. Furthermore, the CCD-camera is aligned with the cascade front when the test section is turned to set higher incidences. Both linear-traversing systems are controlled via serial communication while positioning is set independently for each traversing axis.

To obtain all required temperature informations during the experimental runs, a multitude of thermocouples are installed in the facility components. The coolant conditions are acquired by calibrated thermocouples inside each plenum cavity. Also the temperatures in the mixing chamber along with the temperature conditions in the supply lines are recorded. A thermocouple probe is installed to conduct the inlet temperature of the mainflow. All Keithley multimeter, which are used to record the voltages from the pressure sensors and temperatures, have built-in temperature reference points and a IEEE 488.2 bus connection. Furthermore, the multimeter offer a true 61/2 digit resolution (22-bit). The scan rates for voltage values are up to 200 channels/second with approximately 2000 readings/second for a single-channel scan. The scan rate for temperatures is significantly lower at 2 readings/s.

Calibration of the thermocouples is conducted prior to the experiments. A temperature calibrator was used from Fa. Wika (temperature calibrator Model 9105). It is composed of an aluminium block which is equipped with thermoelectric coolers (TEC). A resistance thermometer Pt-100 is used as sensor and the temperature of the aluminium block is regulated by the TEC. The measurement accuracy of the internal resistance thermometer is ± 0.1°C. The control stability is ± 0.02°C. Due to the poor accuracy of the internal temperature sensor, the thermocouples are calibrated versus a reference thermometer. For this purpose, a second high-precision Pt-100 resistance thermometer with a measurement accuracy of $\pm 0.01\,K$ was used. The calibration of the reference thermometer was carried out by the German Calibration Service DKD. The absolute measurement accuracy for thermocouples implemented in the data acquisition system is then estimated to $\pm 0.2\,K$.

Piezoelectric pressure transducer from Setra Ltd. and Druck Ltd. (type D239 and PMP4100) with different pressure ranges are used to acquire the pressures inside the wind tunnel, the plenum and the coolant supply. A summary of all pressure transducer used for the experiments including their corresponding parameters and absolute measurement uncertainty can be found in the appendix A.5. Pressure transducer with a pressure range up to $7.5\,kPa$ feature a voltage range from 0 ... $5\,V$. For pressure sensors with larger pressure range up to 7 bar, the voltage range is extended to 0 ... $10\,V$. The transducers have an excellent linearity, long-term stability and fast response smaller than $10\,ms$ but they are subject to variation in ambient conditions such as temperature and relative humidity. To determine the transfer function of each sensor, a calibration is conducted using a pressure calibration system from Druck Messtechnik GmbH with an overall measurement accuracy of $0.01\,\%$ full scale. Former tests have shown that the transfer function, which relates the transducer voltage U_T to the measured pressure, is constant. A long-term zero-drift of the sensors occurs but this effect can be reduced and integrated into the measurement system. Optimum stability of the transducers is achieved by applying a constant-temperature environment inside the data-acquisition racks. Furthermore, a daily zero-point calibration for each transducer is carried out. The measured zero-point drift is used to correct the calibration functions.

4 Measurement Techniques

4.1 Pressure Measurement Techniques

4.1.1 Pneumatic Probes

At the wind tunnel, pneumatic probes as pitot and 5-hole probes are used to determine local total pressure losses and the flow field at the cascade inlet and exit. Insertion of probes causes disturbance of the local flow structures. Therefore the probe size must be kept as small as possible to minimize probe interference and displacement effects while taking into account that the probe must still be manufacturable. For the measurements presented here, a pitot probe with a cropped probe head was used which has an inner diameter D_i of $0.8\,mm$. The inner-to-outer diameter ratio of the probe head is $D_i/D_o = 0.66$ and the total length of the probe measures $0.485\,m$. In order to conduct quantitative flow field measurements with pitot-probes in terms of the flow angle, the used probe must be calibrated to determine its angle characteristic. Therefore, calibration is carried out in the calibration channel. For daily service the probe is re-calibrated in the free-jet of the wind tunnel with demounted cascade. The probe is positioned in the center of the exit cross-section and is rotated around its axis between -70° and +70°. The calibration curve is obtained for three different free-stream velocities at $u_m = 44$, 52 and $60\,m/s$. 36 calibration points are acquired with a angle resolution of $\Delta\beta = 5°$.

The angle characteristic of the used probe is shown in Fig. 4.1a). A significant drop of the total pressure in both directions is clearly visible when the probe is turned away from its reference position at $\beta = 0°$. Although increasing the free-stream velocity leads to higher absolute total pressure levels, the angle characteristic is still comparable. Normalizing the total pressure with the free-stream velocity head $p_t - p_0$, leads to collapsing of all calibration points into a single curve. According to this, the freestream velocity for the calibrated range has no effect on the angle characteristic of the used pitot-probe. This conclusion may only be valid for incompressible flows and a limited range of Reynolds-numbers. For highly subsonic or transonic flows, the development of shocks in front of the probe leads to significantly alternated pressure reading. The results of the calibration are stored in files in terms of the angle difference between the left and right pressure slope given as $\Delta\beta_{corr} = 1/2\cdot(\beta_l+\beta_r)$. This value characterizes the asymmetry of the obtained pressure slope with regard to its zero point due to manufacturing tolerances as well as the shape of the probe head (an ideal probe should give values of $\Delta\beta_{corr} \sim 0°$ indicating a fully symmetric angle characteristic).

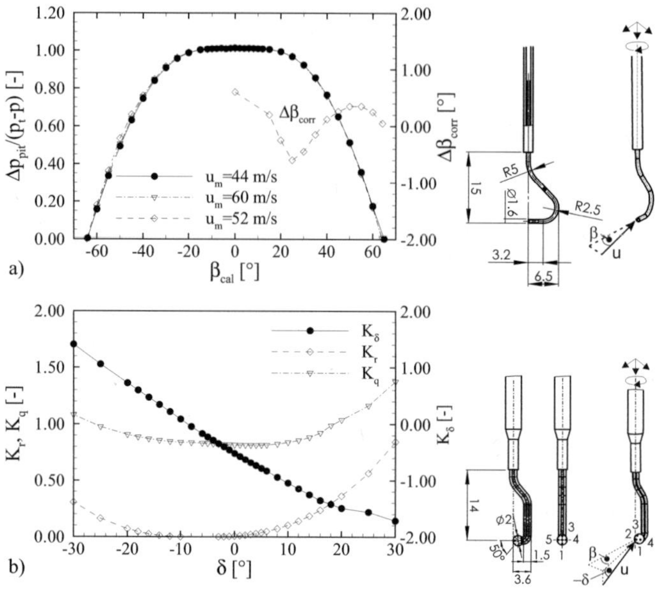

Figure 4.1: Angle characteristics of a Pitot-probe at different free-stream velocities $u_m = 44$, 52 and $60\,m/s$ (a); calibration factors of used 5-hole probe in terms of K_δ, K_r and K_q over the flow angle δ varied between $\pm 30°$ for the maximum freestream velocity $u = 69\,m/s$ (b)

The obvious insensitivity of the probe to changes in total pressure is quite large within a angle range of $\pm 20°$. This can be used to accurately determine the exit flow angle perpendicular to the shaft axis. Thereby, the probe is turned anti-clockwise with a angle offset of $\pm 55°$ - $60°$ into the region of strong pressure gradients to obtain the total pressure at this location. The probe is then turned back to the other side of the pressure slope to match similar magnitudes of the total pressure. Half of the angle difference between the negative and the positive pressure slope represents the flow angle after correcting it with the probe calibration function. The flow angle can then be determined with a measurement accuracy of $\pm 0.1°$. This method is suitable for flow fields with strong gradients of total pressure and has been previously used for measurements of the secondary losses and the cascade exit flow field by Sauer et al. [110] in turbine cascades or by Müller [97] in compressor cascades.

In contrast to pitot-probes where only the flow angle β in the $x-y$ direction can be measured, the use of multiple-hole probes allows the determination of all spatial velocity vector components u, v, w as well as of both flow angles β ($x-y$-plane) and δ (in the $y-z$-plane) simultaneously. Thus enabling the complete description of the three-dimensional flow field downstream the cascade. The used 5-hole probe consists of a spherical head with a diameter of $2\,mm$ and 5 pressure holes. The hole labeled with no. 2 is positioned in the head center in line with the shaft axis. An accurate calibration of the probe is required to determine the influence of the spatial flow direction on the measured pressure at each hole.

The probe calibration is conducted at the calibration channel K-200. It features a circular nozzle outlet section with a diameter of $0.2\,m$. The flow angles β and δ can be set independently and the probe is calibrated at four different free-stream velocities of 16, 30, 50 and $69\,m/s$. To obtain the pressures in the β-direction, the probe is rotated around its shaft axis until the pressure difference between the holes 4 and 5 falls below a defined threshold. Then the flow angle β can be obtained directly. For the δ-direction, the pressure difference between hole 1 and 3 for each calibration point is measured. This requires accurate traversing for a wide range of δ-angles between -30° and +30°. Within the region of small delta angles of $\pm 6°$, the probe is traversed in 1° steps. At larger delta angles, the resolution is slightly coarsened with steps of 2° and 5°, respectively. Finally the obtained pressure differences between the holes are used to obtain the calibration curves. The results are stored in files in terms of following calibration factors

$$K_\delta = \frac{p_1 - p_3}{\Delta p_{24}}, \tag{4.1}$$

$$K_r = \frac{p_{t0} - p_2}{\Delta p_{24}}, \tag{4.2}$$

$$K_q = \frac{p_{t0} - p_0}{\Delta p_{24}}, \tag{4.3}$$

$$with \quad \Delta p_{24} = p_2 - \frac{p_4 + p_5}{2}. \tag{4.4}$$

4.1.2 Wall Pressure Taps

For the determination of the static wall pressures, small holes are implemented into the wall of the tested component. They are connected to pressure sensors by small flexible plastic tubes. This method implies the assumption that the static pressure is constant across the wall boundary layer. Hence, the measured static wall pressure represents the static pressure of the free-stream. The quality of such pressure measurements and their accuracy relies directly on the precision of the tap machining, the hole geometry and the design/length of the tube connection. Investigations on the impact of the hole geometry and the burr height in dependency of the Reynolds number can be found in Nitsche [101].

Here, the error in pressure measurement increases rapidly with increasing dimensionless tap diameter and hole length-to-diameter ratio. Moreover, the measurement error increases also with increasing Reynolds-number but the effect is negligible for subsonic flows. The contamination with dust or occurring secondary flow inside the pressure taps will also affect the static pressure measurement. Reduced influence on the pressure measurement is attributed to the design of the tap exit. Holes with largely angled diffusers can produce differences of 0.3% with respect to the cylindrical reference form and angled or rounded hole edges can alter the pressure reading up to 0.9%.

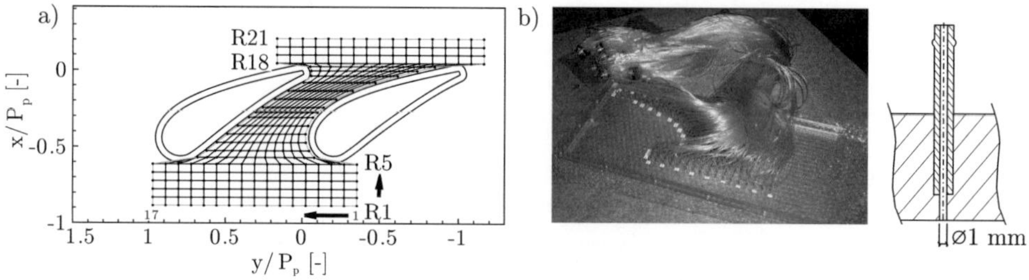

Figure 4.2: Wall tap grid for measurement of the endwall static pressure field (a) and test plate equipped with wall pressure taps and tubulation (b)

Due to these reasons, the tap geometry was carried out as a cylindrical hole with a straight tap exit. The length of the wall tap is selected with $5\,mm$ and the hole diameter is $1\,mm$. Further the holes are expanded to a hole diameter of $1.6\,mm$ which is used to take in the bulged stainless steel tubulation. For sealing and fixing purposes, the tubes are glued into the holes. With the resulting length-to-diameter ratio of 4.54 and the dimensionless hole diameter $d^+ = 55$ (turbulent flow with $u = 15\,m/s$), the measurement error of the wall pressure due to the tap geometry is less than $0.07\,Pa$ according to Nitsche [101] and can be neglected.

The endwall test plate contains 309 wall taps grouped to 21 tap rows, as depicted in Fig. 4.2. The tap rows are labeled, starting from the cascade inlet, from 1 to 21 in streamwise direction. With 13 rows, containing 13 wall pressure taps each, more than half of the endwall taps are located within the passage's center. Furthermore, 5 rows with 85 taps are located in the cascade inlet region and the outlet region is equipped with 4 rows containing 68 wall taps. Each pressure tap row at the cascade's inlet and outlet contains 17 taps with a pitchwise spacing of $\Delta y/P_P = 0.054$. The streamwise row spacing is $\Delta x/P_P = 0.054$.

The profile pressure distributions are determined on the two center vanes in Fig. 4.3a). The results are displayed with regard to the profile coordinate system x_p - y_p shown in Fig. 4.3b). Herein both axes are tangential to the leading edge whereas the x_p - axis has two osculation points on the SS near the trailing as well as the leading edge. The airfoils are instrumented with a total number of 47 wall taps for each profile section. The suction and pressure side is equipped with 17 and 18 wall taps, respectively. The relative distance between two taps on pressure and suction side is 5.3 % with regard to the chord length. Due to the large leading edge radius, 12 wall taps were placed around the leading edge to ensure adequate spatial resolution. Here, the relative distance between two taps reduces to 3.2 %.

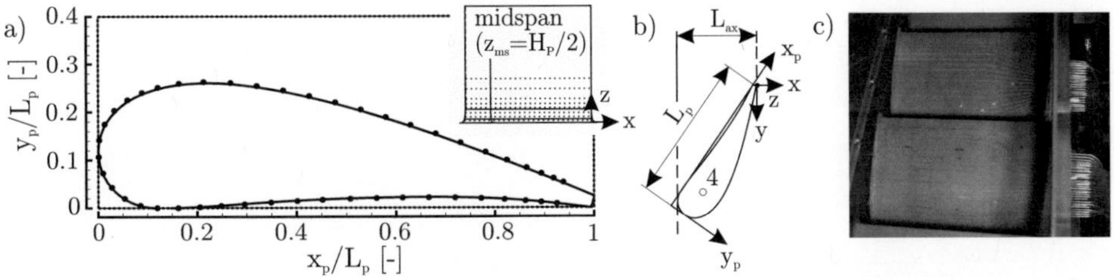

Figure 4.3: Wall tap locations on vane profile (a), definition of the profile coordinate system (b) and instrumented airfoils with wall taps (c)

8 spanwise measurement sections are located near the endwall starting from $z/z_{ms} = 0.033$ up to 0.33, as shown in Fig. 4.3. The spanwise spacing for the first 6 sections is $\Delta z/z_{ms} = 0.033$. For the last two sections closer to airfoil midspan, the distance is increased to 0.066. Small brass tubes with an inner diameter of $1.0\,mm$ are fixed into pre-machined slots of the vane's inlay core. The outer airfoil sleeve is cast around the core and the wall taps with a diameter of $0.5\,mm$ are drilled perpendicular to the surface. The tubes are then connected to the Scanivalve pressure measurement system. In order to measure a specific spanwise pressure distribution, all other taps are closed with highly-adhesive tape. Changing the spanwise position is then achieved by applying the tape to another tap row.

4.2 Flow Rate Measurement

The coolant flow rate is measured by mass flow controllers supplied by Alicat Scientific Ltd. These devices consist of a laminar flow element (LFE) and a proportional pressure valve which is electronically coupled with the flow meter. A schematic of the used flow meters is displayed in Fig. 4.4. The flow rate is set using a PID-controlling loop done by the unit's internal electronics. Its repeatability is $\pm\,0.2\%$ of the full scale range and the unit has a fast response time of $10\,ms$. Each flow meter is pre-calibrated using a high-precision turbine-wheel meter with a measurement accuracy of $\pm\,(0.4\%$ reading $+\ 0.2\%$ of full scale).

Since the LFE consists of small etched plates, it forces the gas in a state of laminar flow beneath the Reynolds threshold of 2000. The pressure drop over the restriction is then measured between its inlet and outlet by a differential pressure sensor. Finally, the viscosity is determined by measuring the temperature at the inlet of the restriction. By selecting the type of gas, the unit provides data for the viscosity as a function of temperature and the compressibility factor. Generally, changes in temperature have tremendous effect on the gas viscosity. Hence, these influences are internally compensated by the unit. The absolute gas pressure is monitored by an absolute pressure sensor with a resolution of maximum $160\,PSIA \pm 0.02\,PSIA$.

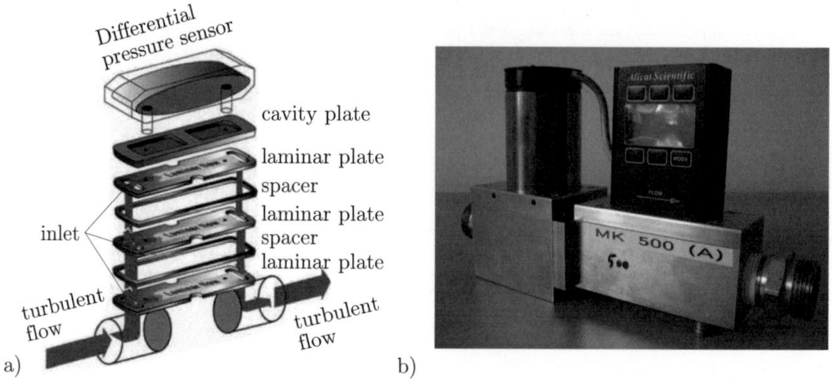

Figure 4.4: Measurement principle of a laminar flow element a) and LFE-flow meter unit with pressure control valve MCR-500SLPM-D from Fa. Alicat Scientific b)

The operating principle of the unit is based on the physics of the Hagen-Poiseuille flow. It quantifies the linear relationship between the pressure drop, flow rate and dynamic viscosity of a gas flow of laminar state inside a pipe. The relationship is known as Poiseuille equation

$$\dot{V} = \frac{\pi R^4}{8\mu L}(p_1 - p_2) = K\left(\frac{\Delta p}{\mu}\right) \tag{4.5}$$

where $\dot{V}$ is the volumetric flow rate, p_1 the static pressure at inlet, p_2 the static pressure at outlet, R the radius of restriction, μ the dynamic viscosity of the fluid and L the length of the restriction. Since π, R and L are constant, the definition is rearranged, such that K is a constant factor determined by the restriction's geometry. The measured flow rate is then only dependent on the pressure drop across the restriction and the gas viscosity. In order to convert the volumetric to a mass flow rate, following definition is used

$$\dot{m} = \rho\dot{V} = \frac{p}{R \cdot T}\dot{V}. \tag{4.6}$$

The density of the measured gas changes with temperature and pressure which must be considered in order to determine the correct mass flow rate. Using ideal-gas laws, the effect of temperature and pressure on density is taken into account with regard to standard conditions as

$$\frac{\rho_a}{\rho_s} = \frac{p_a}{p_s}\frac{T_s}{T_a}, \tag{4.7}$$

wherein the subscript a denotes for the specific value at the actual flow conditions and the subscript s describes the standard condition. If the physical behavior of the gas with changes in temperature and pressure would not follow the ideal gas law, compressibility effects must be considered. The real-gas formulation then introduces the compressibility factor Z as follows

$$\frac{\dot{V}_a}{\dot{V}_s} = \frac{p_s}{p_a}\frac{T_a}{T_s}\frac{Z_a}{Z_s}. \tag{4.8}$$

The flow rate corrections are internally made to 25°C and 14.696 $PSIA$ (101325 Pa) as reference conditions and the gas compressibility factor at these conditions. That allows to calculate the mass flow rate by multiplying the standard mass flow with the real gas density at standard conditions.

4.3 Hot Wire Anemometry

For measuring the main flow turbulence intensity, a CTA-system (CTA ... Constant Temperature Anemometry) from Dantec Ltd. was used in combination with a single sensor probe with a 5 μm platinum-plated tungsten wire (see also Fig. 4.5). The system consists of up to six CTA modules, a calibration unit and a PC with an A/D-converter board. The PC is connected to the measurement system via a serial interface while the analogue output signals from the CTA modules enter the PC via the A/D board. Three frames can be combined to form a 18-channel system.

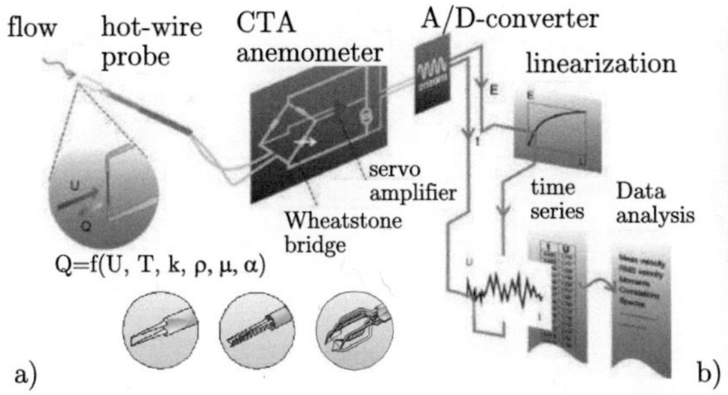

Figure 4.5: a) CTA-measurement principles and b) Streamline CTA-system with calibration unit from Dantec Dynamics (source: www.dantecdynamics.com)

Each mean velocity measurement was obtained from the average of $2 \cdot 10^6$ points at 10000 samples/s. Hence, the sampling rate and measurement time for each boundary layer section was selected with 10 kHz and 20 s, respectively. A temperature probe inside the test rig monitors the flow temperature. The temperature information is used for the correction of the anemometry data.

The hot wire probe was calibrated prior to the measurements to determine its calibration function. Therefore, an external calibration unit from Dantec Ltd. is used. The unit utilizes a system of different nozzles to produce a free-jet flow at different flow velocity ranges. The probes are placed inside the core of the free-jet to determine the relationship between the measured bridge voltage and the velocity magnitude. The pressure difference between the nozzle in- and outlet along with the absolute pressure at the nozzle inlet is obtained by internal pressure transducers. The flow temperature inside the calibrator is monitored by a thermocouple. The isentropic relationship is then used to calculate the flow velocity. Constant flow velocities at the nozzle exit are achieved by controlling the absolute nozzle inlet pressure via an internal pressure-regulating valve. Since the accuracy of the hot wire calibration is mainly dependent on the setting of the nozzle flow-velocity, the behavior of the calibration system was checked versus a Prandtl-probe with a head diameter of $1.8\,mm$ above the nozzle exit.

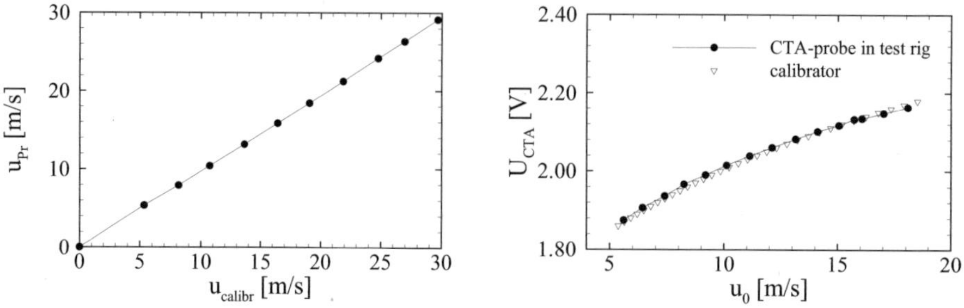

Figure 4.6: Comparison of the free-jet velocity obtained by a Prandtl-probe u_{Pr} with regard to the values which are internally determined by the calibrator u_{calibr} (left) and calibration function of the CTA-voltage signal versus velocity either obtained with the calibrator unit or placing the hot-wire probe in the test duct along with a Prandtl-probe installed closely upstream (right)

The velocity head $p_{dyn} = p_t - p$ and the absolute magnitude of the pressures are acquired with piezo-electric pressure transducers connected to digital multimeter DMM7200 from National Instruments. The obtained pressures were then normalized to standard conditions and the velocities at the Prandtl-probe are calculated. A comparison between the Prandtl-probe and calibrator was conducted within a flow velocity range of 5.3 ... $29.7\,m/s$ taking 10 measurement points. The results are shown in Fig. 4.6 with the velocity estimated from the Prandtl-probe measurements denoted as u_{Pr} and the nozzle-exit velocity u_{calibr} which is internally calculated by the calibrator system. It can be observed that the velocity magnitudes show quite good agreement at low flow velocities. As the pressure difference across the nozzle increases, slight differences of less than $2\,\%$ in the jet velocity are obtained. This may be attributed to the potential flow effect of the Prandtl-probe due to its large head dimension compared to the jet nozzle diameter of $12.4\,mm$ which equals a diameter ratio of 0.145. Furthermore, the velocity calibration of the single hot-wire probe obtained by the calibrator unit is checked by placing the probe into the test rig together with a Prandtl-probe installed closely upstream to measure the flow velocity. The raw voltage signals are acquired within a flow velocity range between 5 ... $18\,m/s$ followed by temperature-correction. Both calibration curves are plotted in Fig. 4.6 and show quite good agreement over the investigated velocity range.

4.4 Thermography using Temperature-Sensitive Paints

Temperature-Sensitive Paint (TSP) is a non-intrusive optical temperature-measurement technique based on thermal-quenched photoluminescence with high spatial resolution of the temperature field. The paint composes of luminescent molecules and a polymer binder material which is non-permeable to oxygen for paint formulations used in temperature measurements. For pressure measurements, the paint composition is selected for oxygen permeability denoted as Pressure-Sensitive Paint (PSP). Both components are dissolved in a solvent which evaporates after the spraying process and leaves behind a polymer matrix with the embedded, immobilized luminescent molecules. The paint is then illuminated by an appropriate light source to excite the molecules which then emit light with longer wavelengths as the excitation light. Both light paths can be separated by properly selected optical filters according to the used paint formulation and illumination. The emitted luminescent intensity is then acquired by either scientific-grade digital cameras or photodiodes. A detailed description of the photo-physical process for both measurement techniques can be found in Liu and Sullivan [84], Sullivan [123], Campbell et al. [24] and Bell [8]. In the past, thermographic phosphors had been proved to be a useful tool to investigate various aerodynamic effects. Global surface heat-transfer distributions on a waverider model and on a 25°/55° cone model at high Ma-numbers in hypersonic wind tunnels were measured by Liu et al. [82] and Hubner et al. [56].

Furthermore, thermal sensitive paints have been used to investigate boundary layer transition in high duration transonic wind tunnels as done by Hammer [49]. The description in the following sections of fundamentals of this measurement technique is focused on TSP but most facts are similar for both techniques. The reader is referred to the book of Liu [84] or to Bell et al. [8] for more detailed informations on the PSP-technique. Recently, the PSP-measurement technique was used more frequently in film-cooling studies, done e.g. by Wagner et al. [130], Charbonnier et al. [26] and Mhetras et al. [96] for turbine blade and platform cooling or by Wright et al. [140] for flat-plate film-cooling investigations. This method takes advantage of the heat-mass-transfer analogy, as described by Jones [59]. Here, film-cooling effectiveness can be obtained by measuring the concentration differences between two cases using different gases. For the first case, air is injected and the partial pressure of oxygen via PSP is measured. In a second case, nitrogen N_2 is blown in instead of air suppressing the luminescent reaction caused by oxygen. It is assumed that since the density of air and nitrogen is very close, the measured static pressure is similar in both cases. The film effectiveness can then be expressed by the ratio of oxygen concentration with air and N_2-injection.

4.4.1 Photophysical Principles

Molecular photoluminescence, which includes phosphorescence and luminescence, is a radiative and reversible process that occurs when a luminescent molecule is promoted to a higher energy level by absorbing a photon of proper excitation light energy. It can return to the ground state by emitting a photon of light or through radiationless intermolecular interaction. As light passes through the active layer, the intensity of light absorbed by the paint is directly proportional to the intensity of luminescence emitted by the paint. While the measurement principle of the TSP-technique is shown in Fig. 4.7 a), the general photo-physical behavior of a

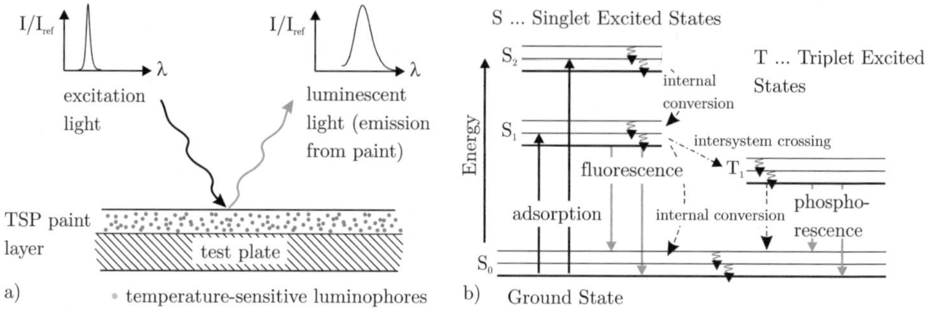

Figure 4.7: Schematic of the TSP measurement technique (a) with Jablonski energy-level diagram (b)

single luminophore during the luminescent process can be described with the Jablonski energy-level diagram, as displayed in Fig. 4.7 b). It is excited from the ground state, denoted as S_0, to an excited electronic state, S_1 or S_2, by absorbing a photon. Thereby, each excited state has different vibrational states and each vibrational state has different rotational states. The excited luminophore returns to the ground state by a combination of radiative and radiationless processes. The light emission from the lowest singlet state is called fluorescence which is a spin-allowed transition between two energy states. For the occurrence of phosphorescence, the excited luminophore is transferred to a triplet state by inter-system crossing. Here, the energy of the first triplet state is typically lower than the corresponding singlet state. Phosphorescent light is than emitted from the paint by returning to the ground state S_0 which is a spin-forbidden transition between states of different multiplicity. Therefore, the fluorescent lifetime is much shorter than the process of phosphorescence. Furthermore, the excited molecule can pass down to the ground state without emitting luminescent light. This process is called radiationless deactivation which involves electronic processes as internal conversion as well as inter-system crossing and external conversion. Internal conversion appears to be efficient when two energy levels are close to each other. For more information on the general principles of luminescence, the reader is referred to Liu and Sullivan [84].

The emitted luminescent light is changed by various quenching processes with thermal quenching which is related to TSP and oxygen quenching which is used in PSP-measurements. Since not all of the excitation light is transferred to the emission process, the light emitted from the paint has a lower energy as the excitation light. Thus, the shift of the emitted light to longer wavelengths (Stokes shift) allows to separate both light parts by optical filters. For TSP, it is assumed that the paint layer is not permeable to oxygen, so that only the thermal quenching (radiation-less deactivation) affects the luminescence emission of the TSP. As the temperature increases, the probability of radiation-less deactivation is much higher. Thus, it reduces the luminescent signal emitted from the paint and the paint appears to be darker on the acquired image with increasing temperature. Three different images are required to use this technique for quantitative temperature measurements. At first, a dark/offset image is acquired (with no flow, no light) which includes the thermally induced dark current in the CCD-sensor and the intensity offset of the chip set by the camera electronics. Further, a wind-off or reference image (with no flow, lights on) is acquired which in this case refers to a state of main flow on and no coolant injection. At last, a wind-on image (with flow, coolant injection and lights on) is taken. A series of 20 images is acquired for each image type.

In order to quantify those processes, the ratio of the rate of luminescence emission divided by the rate of excitation is introduced. The so-called quantum yield is described by the following equation

$$\Phi = \frac{rate\ of\ luminescence\ emission}{rate\ of\ excitation} = \frac{I_e}{I_a}. \tag{4.9}$$

The Lambert-Beer-law describes the weakening of light intensity depending on the traveled distance when it passes through an absorbing substance in conjunction with the intensity attenuation correlated by the concentration of the absorbing substance. Following the law, the absorption intensity can be calculated by

$$I_a = I_{ex} e^{(-\alpha_M c L)}, \tag{4.10}$$

where α_M is the molar extinction coefficient, L the optical path length traversed by the light beam through the substance and c the concentration of the absorbing species.

For the usage of TSP in an experiment, the assumption is made that the binder is not permeable to oxygen, so that equation (4.9) can be simply rewritten as

$$\Phi = \frac{I}{I_a} = \frac{k_L}{k_L + k_D} \tag{4.11}$$

where k_L is the rate of reaction for luminescence emission and k_D the rate for radiation-less deactivation which may be composed of a temperature-independent term k_0 and a temperature-dependent term $k_0(T)$. The temperature-dependent part $k_0(T)$ can be assumed to follow an exponential form which was found by Arrhenius [113]. The Arrhenius equation describes the temperature dependency of chemical reaction rates k as

$$k_0(T) = A e^{-\frac{E_A}{R_u T}}, \tag{4.12}$$

where A is a constant factor, R_u the universal gas constant, T the temperature and E_A the Arrhenius activation energy. The activation energy is equal to the energy that is required to excite and traverse a pair of electrons in a molecule from the ground state to a higher energy level. Following this, the relation for the quantum yield can be rewritten by considering the temperature dependency of the radiationless deactivation rate

$$\frac{1}{\Phi(T)} - \frac{1}{\Phi(T=0)} = \frac{I_a}{I(T)} - \frac{I_a}{I(T=0)} = \frac{k_L + k_0 + k_0(T)}{k_L} - \frac{k_L + k_0 + k_0(T=0)}{k_L} \tag{4.13}$$

$$\frac{I_a}{I(T)} - \frac{I_a}{I(T=0)} = \frac{1}{k_L} k_0(T) = \frac{1}{k_L} A \exp\left(-\frac{E_A}{R_u T}\right), \tag{4.14}$$

where $I(T=0)$ is the luminescence intensity at absolute zero. The absorption intensity I_a can be eliminated by dividing the equation given above with a reference intensity I_{ref} at a known constant temperature

$$ln\left[\frac{I\left[I(T=0) - I_{ref}\right]}{I_{ref}\left[I(T=0) - I\right]}\right] = \frac{E_A}{R_u}\left(\frac{1}{T} - \frac{1}{T_{ref}}\right). \tag{4.15}$$

Under laboratory conditions, the reference temperature T_{ref} is close to the temperature where the intensity is measured. Therefore, the factor $[I(T=0) - I_{ref}]/[I(T=0) - I)]$ is nearly unity. Thus, equation (4.16) can be simplified and written as

$$ln\left[\frac{I(T)}{I_{ref}(T_{ref})}\right] = \frac{E_A}{R_u}\left(\frac{1}{T} - \frac{1}{T_{ref}}\right) \tag{4.16}$$

wherein T and T_{ref} are the temperatures at wind-on and reference condition, respectively. I_{ref} is the luminescent intensity obtained at the reference temperature T_{ref}. According to Liu and Sullivan [84], the Arrhenius equation can accurately represent the temperature dependency over a certain temperature range for a wide class of TSP's. For this, equation (4.16) can be rearranged and expressed in an equivalent exponential form

$$I = I_{ref} \exp\left[\frac{1}{K_t}(T_{ref} - T)\right], \tag{4.17}$$

$$K_t = \frac{R_u T T_{ref}}{E_A}, \tag{4.18}$$

where K_t is a constant factor which depends on the paint formulation used for a certain temperature range. The factor K_t of a paint formulation can be determined by a temperature calibration. Below 0°C the temperature dependency differs significantly from the Arrhenius-plot as the luminescence reaction of the molecules to the excitation decreases. Cattafesta and Liu [25] have shown that K_t is related to the normalized paint sensitivity. Smaller values of K_t for a specific paint increase the paint sensitivity and decrease the measurement uncertainty.

4.4.2 Setup of Measurement Technique

In general, the thermography measurement system using temperature-sensitive paints consists of a photo detector, the illumination system, the optical filtering and the data processing unit.

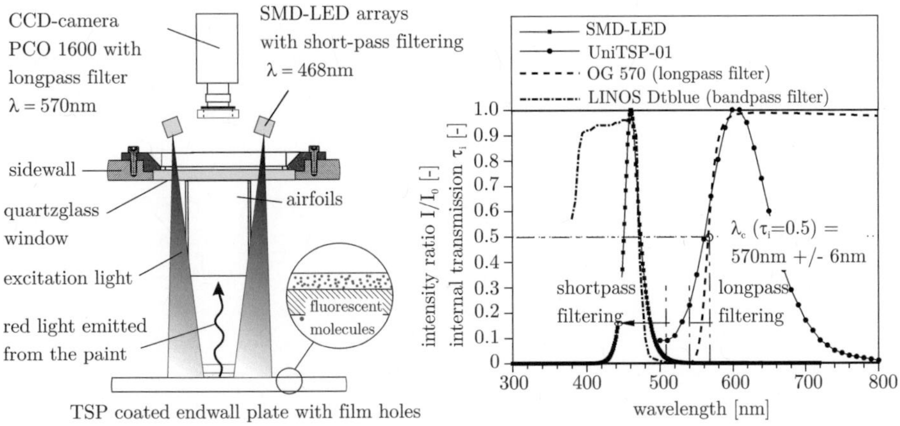

Figure 4.8: Schematic of the TSP setup for cascade testing along with the emission wavelength spectrum of the used LED & TSP-paint (reproduced with permission of Innssi, Inc.) and the internal transmission curves for the used short- and long-pass filtering (reproduced with permission of Linos Photonics GmbH & Co. KG)

For intensity-based systems, the common method is to use a CCD-camera system as photo detector for the luminescence. For these experiments, the luminescence intensity is detected by a scientific-grade CCD camera PCO-1600 that provides high intensity resolution of 14-bit and high spatial resolution of 1600 x 1200 $pixel$. The embedded CCD-chip has a sensor format of 12.2 mm x 9 mm. The pixel size is 7.4 μm x 7.4 μm with a full well capacity of 40000 e^- per pixel. Thermoelectric cooling of the CCD-sensor is applied with a maximum temperature decrease of 50 K below the ambient conditions. It reduces the thermally induced dark current to 0.01 e^-/pixel·s at a typical operating temperature level of the chip of -20°C. The camera is connected to the image-storing and data-processing PC-system via an IEEE-1394 interface and fiber-optic cables.

The CCD-camera is equipped with a long pass filter placed in front of the sensor to cut-off the blue light emitted by the illumination source. The filter utilizes a cut-on length of $\lambda_c = 570\,nm \pm 5\,nm$ which is defined at 50 % of the maximum internal transmission intensity. But long-pass filtering starts much earlier at a wavelength of 540 nm. The wavelength spectra for the used filter is shown in Fig. 4.8. Due to the overlap between the TSP emission curve and the wavelength spectrum around 500 nm - 510 nm which is typically emitted from the LED, some portion of the illumination light still passes through the long-pass filter. Therefore, plane short-pass filters are installed in front of the LED-lamps to avoid this spectral leakage. Filters from Linos Photonics GmbH & Co. KG with a cut-off wavelength at about 502 nm are used. This closes the spectral gap by cutting off all illumination light at larger wavelengths directly after emission. Several Nikon Nikkor lenses with different focal lengths (f = 35, 50, 58 and 85 mm) are available. All experimental results reported here are obtained using the lens with a focal length of $f = 58\,mm$ due to its excellent photo sensitivity of $f/1.2$. It is defined as the reciprocal of the smallest aperture value (f-stop) and expresses the light transmissivity of an optical system at its largest aperture.

For the film-cooling experiments on the flat plate, the painted test surface is illuminated by a pair of LED-strips with 27 blue SMD-LED. The strips are connected to the power supply to control the illumination intensity of the light-emitting diodes. Due to the larger endwall test area that has to be illuminated for the film-cooling experiments, the number of LED-strips of the system was extended. It contains 8 strips with 27 blue LED each to cover the entire region of the film-cooled endwall. The connected power supply is designed to control and adjust the LED-power for single as well as multiple strips or the entire LED-field. Therefore, non-uniform illumination of the endwall can be reduced. The used system is shown in Fig. 4.9.

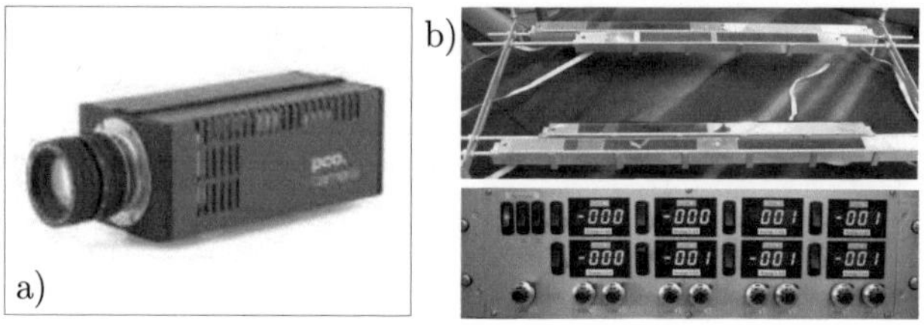

Figure 4.9: Camera-system PCO 1600 (a) and LED - illumination system with power supply (b)

A monochromator/spectrograph system "Cornerstone 260" from Oriel Instruments with a blazed diffraction grating was used to obtain the emission characteristics of the LED-light sources. The emission light of the LED-bulbs is focused onto the entrance slit of the monochromator system by a cylinder lens. The intensity of the exiting light diffracted at the internal grating is then measured as a function of the wavelength. The obtained raw data is corrected with the efficiency curve of the diffraction grating. Observing the results displayed in Fig. 4.8, it appears that the LED-bulb has its emission peak at a wavelength of $462\,nm$. The wavelength band is in the range of $\pm\,10\,nm$ being defined at $50\,\%$ normalized intensity. At smaller illumination intensities, the range of emitted wavelengths is significantly broadened up to $\pm\,50\,nm$. The long-term instability of the excitation light source is less than $1\,\%$ after $20\,min$ of operation.

All experiments reported in this thesis are carried out using a commercially available paint UniTSP-01 from Innssi (Innovative Scientific Solutions, Inc.) with a working temperature range of $293\,K < T_w < 363\,K$.

4.4.3 Calibration

In order to quantitatively measure film effectiveness values, a calibration of the used TSP coating is needed in order to relate the measured luminescence intensity to temperature. The apparatus used for calibration is based on the equipment described in Liu et al. [81] and Sullivan [123]. The TSP is applied to an aluminum plate ($40\,mm$ x $40\,mm$ x $1\,mm$) with 5 calibrated K-type thermocouples implemented into holes of $0.5\,mm$ diameter close to the painted surface. A Peltier thermoelectric heater is connected to the plate via high thermal-conductivity grease and is used to control the block temperature. Further, it is thermally anchored to a copper bar which rests on an axis aligning the calibration sample parallel to the CCD-sensor. The TSP sample is then illuminated by the excitation light source. The illumination intensity is adjusted so that the CCD-sensor saturation is about 75 % of the maximum dynamic range (for 14-bit A/D converter the saturation limit is 16384 gray-scale levels).

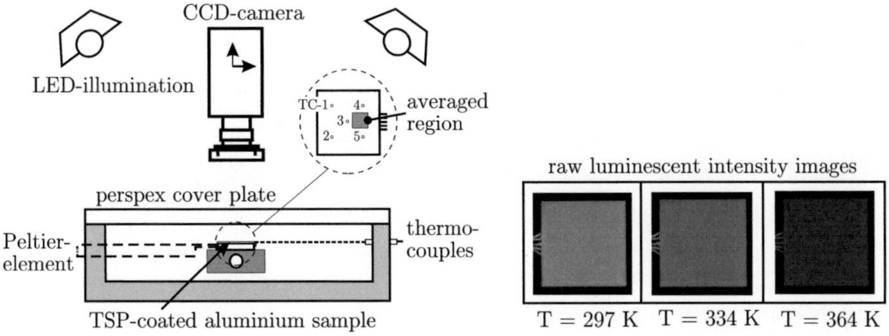

Figure 4.10: Schematic of the TSP-calibration system

17 measurement points were acquired within the TSP temperature range of $283\,K$ up to $363\,K$. Different temperatures are set by changing voltage supplied to the Peltier element within a range from 0 ... $1.9\,V$. Temperature stability of $\pm\,0.1\,K$ is reached after $15\,min$ of operation for each calibration point. The aluminum plate coated with TSP is then illuminated by the LED-light source. The luminescent intensity is acquired by the CCD-camera. Three different series with 20 images each are obtained with an imaging frequency of 30 frames/s. First, an offset image series is acquired with the camera shutter closed and the illumination lights switched off. The reference image series is taken at ambient temperature with the shutter open and the illumination light switched on. The measurement images are then acquired at the corresponding temperatures after the thermal equilibrium was achieved. The images were then averaged on pixel-by pixel basis and the averaged offset image is subtracted from each intensity image. The intensity ratio I/I_{ref} was determined for each pixel of the averaged images for

reference and the measurement condition following

$$\frac{I}{I_{ref}} = \frac{I - I_{offs}}{I_{ref} - I_{offs}}. \tag{4.19}$$

The luminescence intensity distribution is analyzed in a region of 100 *pixel* x 100 *pixel* for each calibration point to determine the spatially averaged intensity ratio I/I_{ref} versus temperature. This region is placed close to the center thermocouple denoted as TC-3 in Fig. 4.10 to avoid areas of non-uniform illumination and luminophore concentration on the test plate. The data is post-processed by a statistical analysis on a representative sample of 10000 *pixel* using the probabilistic tool ProSi-Post from Voigt [129]. It is performed to calculate mean values and standard deviation of the intensity ratio. The temperature measured by the thermocouples is related to the mean values of the intensity ratio I/I_{ref}. The temperature deviation between the thermocouples on the aluminum plate is less than ± 0.2 K.

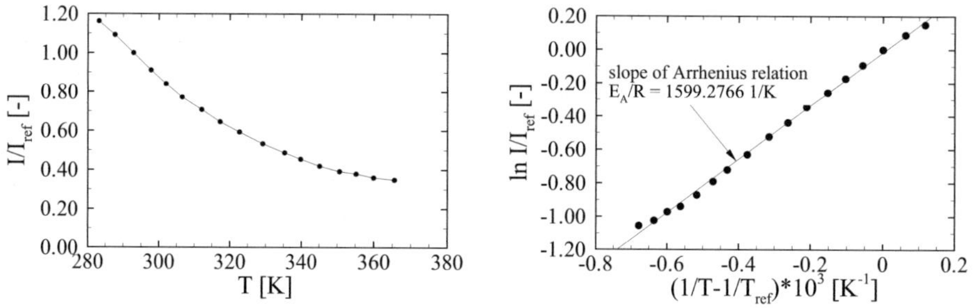

Figure 4.11: Luminescence intensity calibration of UniTSP-01 (left) within a temperature range 280 K - 360 K along with the corresponding Arrhenius-plot (right)

The obtained calibration for UniTSP-01 is shown in Fig. 4.11 in terms of the luminescent intensity ratio versus temperature. Observing the calibration results, it is obvious that the luminescent intensity decreases steadily with increasing temperature due to the optical quenching process. It weakens the fluorescence emitted by the luminophores that are embedded in the paint. An overall reduction of about 60 % is observed over the working temperature range. Less change in luminescent intensity is observed for the temperature approaching the upper and lower limit of the temperature range.

The right diagram in Fig. 4.11 shows $\ln(I/I_{ref})$ versus $1/T$ also denoted as Arrhenius plot. Considering equation (4.16) it gives a straight slope of E_A/R_u. Plotting the calibration data into the diagram it can be observed that the data for UniTSP-01 fits the Arrhenius relation given in equation (4.16) within its working temperature range.

4.4.4 Image Data Processing

The primary task of the image-processing is to calculate temperature distributions from the acquired TSP-images using the light-intensity information emitted from the luminescent paint. Therefore, an in-house software tool was developed to process and analyze the large amount of TSP-data. Here, several tasks have to be fulfilled:

- averaging of image series on a pixel-by-pixel base
- reduction of dark current and compensation of the electric offset for the CCD-sensor
- rationing of both averaged images: wind-on/wind-off
- calculation of surface temperatures using TSP-calibration data
- mapping of temperature images to spatial coordinates

Different images are required for the TSP-technique whereas the ratio between an image at reference temperature and the actual test image is taken. The ratio between the two luminescent intensity values is then used to calculate the surface temperature. During the post-processing a pixel-by-pixel averaging is used to reduce the random noise like photon shot noise. This helps to increase the signal-to-noise ratio (SNR) for each image series as it was proposed by Mendoza [94]. Prior to each test run, a dark image series is acquired by closing the shutter

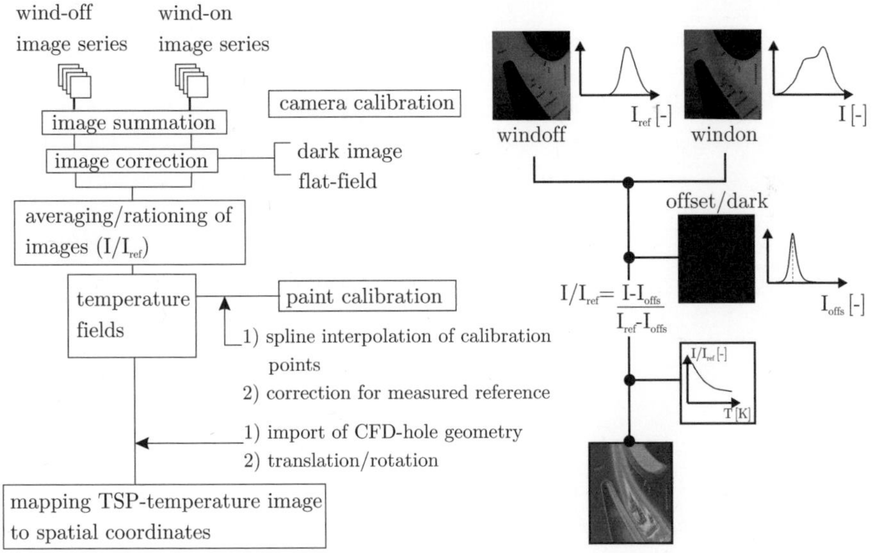

Figure 4.12: Schematic of image processing procedure for TSP-measurements

of the camera and switching off the illumination light. Further, the reference image is taken with the wind tunnel running at its operating point and no coolant injection. This avoids misalignment of cascade components due to thermal expansion. At last, the final test image is taken with coolant injection at higher temperature than the main flow and the illumination light switched on. The aperture time is maintained for all acquired images.

The ratio of the noise-corrected data images can not be taken if the wind-on and wind-off images may show misalignment due to test model deformation under aerodynamic loads. Rationing non-aligned images may lead to considerable errors. Aerodynamic flow phenomena like shocks, boundary layer transition or flow separation or any other geometrical structure can be smeared in the image. Therefore, misalignment between the wind-on and wind-off images was estimated to be approximately one pixel. Thus, the application of image registration methods is not required according to Cattafesta et al. [25].

The wall temperature distribution is calculated based on the temperature function of the TSP which was found during a-priori calibration. Therefore, the calibration data is imported into the image-processing tool and corrected for the actual reference temperature. The surface temperature for each luminescent intensity ratio I/I_{ref} is found by spline interpolation between the corresponding calibration points. This process is repeated for each image pixel.

In order to link the acquired images to spatial coordinates, the scaling factor from pixel to real space was derived by observing marker objects on the endwall surface. The marker position with regard to the used airfoil coordinate system is accurately placed by CNC-machining within a tolerance of $\pm 0.01\,mm$. Since only the marker is coated with TSP, its edges can then be iteratively determined at the point where the gradient of luminescent intensity shows a sudden drop. With the known dimensions of the marker, the image scaling can be performed. After that, the temperature field is transferred to the airfoil coordinate system by translation and rotation of the image. To ensure correct image transformation, the hole locations are checked by comparing the hole exit contour visible in the TSP-image and the point coordinates of the location where the hole axis emerges through the endwall to the design definition. The image transformation is manually fine-tuned until the center points of the hole exits match with the image contours.

5 Data Reduction Methods

5.1 Analysis of Mainflow

Measurements of the spatial total pressure loss and the flow field are carried out at the cascade inlet to determine the inlet boundary layer and at the cascade exit. The measured values are normalized to standard conditions on order to eliminate the effect of varying ambient conditions due to the open-cycle operation of the LSRCW. Standardization of the pressures is carried out with

$$\Delta p_t = p_t - p_{amb,M} \tag{5.1}$$

$$\Delta p_{t,n} = \Delta p_t \frac{p_s}{p_{amb,M}} \tag{5.2}$$

$$\Delta p_n = \Delta p \frac{p_s}{p_{amb,M}} \tag{5.3}$$

using standard conditions of $p_s = 101325\,Pa$ and $T_s = 288.15\,K$ according to SI-units.

As the pitchwise and radial pressure gradients in the cascade represent the driving forces for the secondary flow, the static pressure distribution at the endwall and around the profiles is of great interest especially at increased loading conditions. Furthermore, the endwall static pressures are used to evaluate the pressure gradient across each film hole and hence to quantify the coolant distribution across a specific row or field of film holes. Therefore, measurements of the static pressures are done on the center endwall and the enclosing airfoils. The results are given in terms of the dimensionless pressure coefficient with respect to the velocity head at the cascade exit $p_{dyn,2} = p_{t2} - p_2$ given by following equation

$$C_{p2} = \frac{p_x - p_2}{p_{t2} - p_2}. \tag{5.4}$$

Further, the isentropic Mach number is used for interpretation and comparison of the endwall static wall pressures with the numerical results. It is defined with

$$Ma_{is} = \sqrt{\frac{2}{\kappa - 1}\left[\left(\frac{p_{t0}}{p_x}\right)^{\frac{\kappa-1}{\kappa}} - 1\right]} \tag{5.5}$$

where p_{t0}/p_x is the total-to-wall pressure ratio. Flow field measurements are conducted at the cascade inlet and exit with the positions of the measurement planes described in detail in chapter 3.2.1. Total pressures and axial flow angles were measured with a pneumatic probe.

Furthermore, 5-hole probes are used to determine the streamwise vorticity (SVO) and the velocity vectors to characterize the secondary flow. Based on the acquired data, calculation of the flow field variables like velocity, flow angles and total pressure loss coefficients were performed. Their definitions are summarized during the next section. Measurements were taken across two passages with 19 traces distributed in spanwise direction to check for flow symmetry. Using this measurement technique, only time mean values are acquired.

The boundary layer edge of the freestream is described by the wall distance where the local velocity equals 99 % of the mainflow velocity in the non-viscous flow region. This quantity is denoted as the boundary layer thickness δ_{99}. Since this definition has a mostly random character and lacks of comparability for different boundary layers, further parameters are determined to conveniently describe the boundary layer. The displacement thickness δ^* of the boundary layer assuming an incompressible flow is determined using

$$\delta^* = \int\limits_{z=0}^{\infty} \left(1 - \frac{u(z)}{u_\infty}\right) dz \tag{5.6}$$

where u_∞ is the velocity at the boundary layer edge. The momentum thickness θ can be calculated by

$$\theta = \int\limits_{z=0}^{\infty} \frac{u(z)}{u_\infty} \left(1 - \frac{u(z)}{u_\infty}\right) dz. \tag{5.7}$$

The shape factor describes the ratio of momentum-to-displacement thickness and can be used to evaluate the quantitative nature of a boundary layer. It is derived from

$$H = \frac{\delta^*}{\theta}. \tag{5.8}$$

The local total pressure loss coefficient is determined as

$$\zeta_{i,l}(y,z) = \frac{p_{t0} - p_{ti}(y,z)}{p_{t0} - p_2}. \tag{5.9}$$

It is then reduced by applying the following area-averaging method

$$\overline{\overline{\zeta}}_{i,a} = \frac{2}{H_P P_P} \int\limits_{z=0}^{\frac{H_P}{2}} \int\limits_{y=0}^{P_P} \zeta_{i,l}(y,z)\, dy dz \tag{5.10}$$

or mass flow-averaging method

$$\overline{\overline{\zeta}}_{i,m} = \frac{\int_0^{\frac{H_P}{2}} \overline{\zeta}_{i,m}(z)\, \overline{\rho u_{ax}}(z)\, dz}{\int_0^{\frac{H_P}{2}} \overline{\rho u_{ax}}(z)\, dz} \tag{5.11}$$

with

$$\overline{\zeta}_{i,m}(z) = \frac{\int_0^{P_P} \zeta_{i,l}(y,z)\,\rho u_{ax}(y,z)\,dy}{\int_0^{P_P} \rho u_{ax}(y,z)\,dy} \tag{5.12}$$

where i refers to the measurement position: $i = 1$ the inlet plane (MP1) and $i = 2, 3, 4$ identifies the plane downstream of the cascade (see also Fig. 3.5 for positions), P_p is the pitch and H_p the blade height. The net endwall loss is calculated by subtracting the averaged loss of the inlet boundary layer and the profile loss at midspan. The latter is obtained by averaging the measured total pressure loss at three sections around the blade midspan. As long as the endwall loss areas do not merge over the blade, the result is independent of the blade span when the profile loss is subtracted. It presents the isolated secondary loss. Hence, it can be expressed with respect to a reference blade height $H_{ref} = 0.1\,m$.

$$\zeta_{ew} = (\overline{\overline{\zeta}}_2 - \overline{\overline{\zeta}}_{1,ibl} - \overline{\overline{\zeta}}_P)\frac{H_P}{H_{ref}}. \tag{5.13}$$

To investigate the development of the secondary flow structures downstream of the cascade, the vector field of the secondary flow velocities is of great interest. Therefore, an orthogonal coordinate system is introduced (see Fig. 5.1) for each traversing point which is orientated on the primary two-dimensional flow at vane midspan. Due to that, the x_S-axis is parallel to the midspan flow angle β_{ms} and y_S is the orthogonal axis. The z_S - axis is parallel to the blade axis. With this aligned coordinate system, the projection of the spatial velocity vector $u\,(y,z)$ into the $x_S - y_S$ plane delivers the velocity components of the secondary flow vector by

$$u_{S,ys} = \sqrt{(u_x{}^2 + u_y{}^2)}sin(\beta(y,z) - \beta_{ms}(y)), \tag{5.14}$$

$$u_{S,zs} = u_z, \tag{5.15}$$

$$u_S = \sqrt{(u_{S,ys}^2 + u_{S,zs}^2)}. \tag{5.16}$$

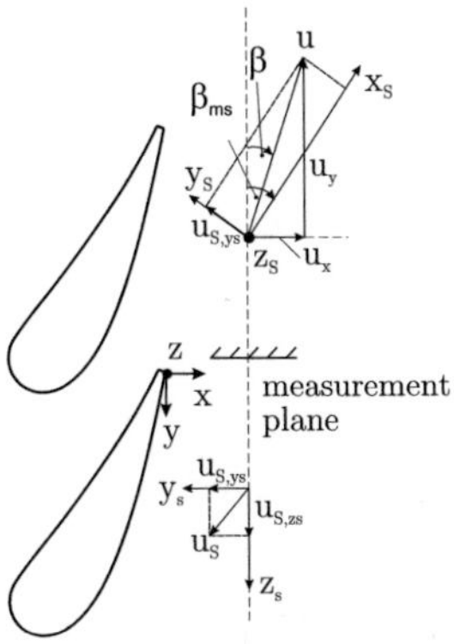

Figure 5.1: Definition of secondary flow vectors

The resulting secondary-flow vector u_S is then transferred back to the measurement plane. It represents the deviation of the spatial velocity from the flow velocity at midspan. For quantitative assessment of the secondary vortices and localization of the vortex center position, the calculation of the vorticity in streamwise direction can be used as it was introduced by Gregory-Smith [46]. This method was successfully used by Weiß [134], Duden [30] or Sauer [110] to quantify the strength of the rotation of the secondary flow.

The vorticity in streamwise direction is calculated from the velocity gradients in the specified measurement plane following the definition of the vorticity as rotation of the vector field of the velocity using

$$\mathbf{\Gamma} = rot\mathbf{u} \tag{5.17}$$

along with the definition of the vorticity components in axial, pitchwise and spanwise direction for the used coordinate system $x - y - z$ as

$$\Gamma_x = \frac{\partial u_z}{\partial y} - \frac{\partial u_y}{\partial z}, \tag{5.18}$$

$$\Gamma_y = \frac{\partial u_x}{\partial z} - \frac{\partial u_z}{\partial x}, \tag{5.19}$$

$$\Gamma_z = \frac{\partial u_y}{\partial x} - \frac{\partial u_x}{\partial y}. \tag{5.20}$$

$$\tag{5.21}$$

Since the streamwise vorticity is defined as the component of the vorticity vector whose axis is parallel to the velocity vector at midspan, it can be calculated with the vorticity component of Γ_x and Γ_y tangential to the midspan flow direction along the pitchwise direction by

$$\Gamma_S = \Gamma_x sin(\beta_{2,ms}(y)) + \Gamma_y cos(\beta_{2,ms}(y)). \tag{5.22}$$

The derivative of the normal velocity components in axial direction, as given in (5.19) and (5.20), can not be determined by interpolation of the data between two measurement planes. This is due to the large distance between the planes that would neglect changes in the velocity gradient in flow direction. Therefore Gregory-Smith [46] introduced an approach of calculating the vorticity in pitchwise direction without to know the derivative in axial direction which is only valid for incompressible flows. Niehuis et al. [100] applied a different method that uses the momentum equation for steady-state, compressible and viscous flows

$$\mathbf{u} \cdot grad\mathbf{u} = -\frac{1}{\rho} grad\, p + \nu \Delta \mathbf{u}. \tag{5.23}$$

The equation can be simplified by assuming high Reynolds numbers in the free-stream. This neglects the viscous term outside the boundary layers. This leads to Euler's equation of motion

$$\mathbf{u} \cdot grad\mathbf{u} + \frac{1}{\rho} grad\, p = 0. \tag{5.24}$$

Following the identity

$$\mathbf{u} \cdot grad\mathbf{u} = grad\, \frac{u^2}{2} + rot\mathbf{u} \times \mathbf{u} \tag{5.25}$$

and combining equation (5.24) with the definition of the vorticity as given in equation (5.17) results in

$$\mathbf{\Gamma} \times \mathbf{u} = -\frac{1}{\rho} grad\, p - grad\, \frac{\mathbf{u}^2}{2}. \tag{5.26}$$

Thus, the vorticity in pitchwise direction Γ_y can be calculated with

$$\Gamma_y = \frac{1}{u_x}\left[\frac{1}{2}\frac{\partial u^2}{\partial z} + \frac{1}{\rho}\frac{\partial p}{\partial z} + u_y \Gamma_x\right]. \tag{5.27}$$

Using equation (5.22), (5.18) and (5.27), the vorticity in streamwise direction Γ_S can be calculated. The normalized parameter, denoted as SVO, is hereby referred to the profile length and the isentropic velocity at the cascade exit $u_{2,is} = p_{t1} - p_2$ following

$$SVO = \frac{\Gamma_S L_P}{u_{2,is}}. \tag{5.28}$$

5.2 Analysis of Coolant Flow

Taking the present endwall cooling scheme, the shaped film holes are distributed across the entire endwall of the turbine passage to cover as much endwall surface area with coolant as possible. Due to strong acceleration of the mainflow fluid through the turbine blade passage, the main flow static pressure field varies in streamwise and pitchwise direction. Furthermore the endwall pressure field is altered by increased airfoil loading, strongly influencing the coolant distribution of each film row. With this the varying endwall pressures alter the hole pressure ratio and hence the hole flow rate. The pressure inside the supply plenum is considered to be constant. To characterize the strength of injection for a specific row, it is useful to obtain an averaged blowing rate by

$$M = \frac{\rho_c u_c}{\rho_m u_m} = \frac{\frac{\dot{m}_c}{A_h}}{\rho_m u_m} \tag{5.29}$$

where the averaged mass velocity $\rho_m u_m$ is obtained from the endwall static pressure measurements, $\dot{m}_c$ is the coolant mass flow rate supplied to the film row and A_h is the sum of all hole cross sections. This parameter is maintained for all incidence cases. Evaluating the variation of the blowing rate across a film row for a given test case, the change in main flow mass velocity has to be taken into account with higher loading condition. The local hole blowing rate is defined with the mass velocity of the main flow at the each hole injection location as

$$M_i = \frac{\frac{\dot{m}_{c,i}}{A_{h,i}}}{\rho_{m,i} u_{m,i}}. \tag{5.30}$$

Hence, the momentum flux ratio per hole can be expressed as

$$I_i = \frac{\rho_{c,i} u_{c,i}^2}{\rho_{m,i} u_{m,i}^2} = M_i^2 \frac{1}{D_{R,i}} \tag{5.31}$$

by accounting for the local blowing rate and coolant density ratio for each film hole. The main flow mass velocity, denoted as $\rho_{m,i} u_{m,i}$, at the hole injection location i can be determined from the endwall static pressure measurements. For this, a non-viscous flow is assumed which leads to a constant total pressure across the passage with $p_{tm,i} = p_{t1}$ and adiabatic conditions with $T_{tm,i} = T_{t1}$ by

$$\rho_{m,i} u_{m,i} = \frac{p_{t1}}{R_m T_{t1}} \left(\frac{p_{t1}}{p_{ew}}\right)^{-\frac{1}{\kappa_m}} \sqrt{\left(\frac{2\kappa_m R_m T_{t1}}{\kappa_m - 1}\right)\left[1 - \left(\frac{p_{t1}}{p_{ew,i}}\right)^{\frac{\kappa_m - 1}{\kappa_m}}\right]}. \tag{5.32}$$

Following this, the individual mass flow rate per hole is computed with

$$\dot{m}_{c,i} = C_D \, \dot{m}_{th,i} \tag{5.33}$$

with the parameter C_D describing the discharge coefficient of a specific film row. The theoretical mass flow rate $\dot{m}_{th,i}$ is defined as

$$\dot{m}_{th,i} = \rho_{c,i} u_{c,i} A_h = \frac{p_{tc}}{R_c T_{tc}} \left(\frac{p_{tc}}{p_{ew,i}} \right)^{-\frac{1}{\kappa_c}} \frac{\pi}{4} D^2 u_{c,i} \tag{5.34}$$

with the isentropic coolant velocity

$$u_{c,i} = \sqrt{ \left(\frac{2\kappa_c R_c T_{tc}}{\kappa_c - 1} \right) \left[1 - \left(\frac{p_{tc}}{p_{ew,i}} \right)^{\frac{\kappa_c - 1}{\kappa_c}} \right] }. \tag{5.35}$$

This definition assumes an isentropic, one-dimensional expansion across the hole with p_{tc}/p_{ew} the driving coolant pressure ratio per hole for the coolant injection. Assuming very small velocities inside the plenum cavities, the measured total coolant quantities can be replaced by their static values which applies to the temperature $T_{t,pl} = T_{pl}$ and to the pressure $p_{t,pl} = p_{pl}$. To finally obtain the hole blowing rates, the physical properties of the coolant and mainstream fluid have to be considered. These parameters are satisfactory described by the specific gas constant R and the adiabatic exponent κ for pure, ideal gases. To control the coolant density, hence the density ratio, mixing of pure gases with different molecular weight can be applied. This would significantly alter the physical properties of the mixed coolant. For this purpose, the adiabatic exponent $\kappa_{c,mix}$ and the specific gas constant $R_{c,mix}$ have to be evaluated as function of the mixed mass fraction.

5.3 Evaluation of Streamwise Film Effectiveness

Film effectiveness distributions are obtained from the raw surface temperature images $T_w(x, y)$ as outcome of the image post processing applying equation (2.6). Due to the incompressible state of the main flow, the recovery temperature is replaced by the static gas temperature. Since an endwall material with small but finite heat conductivity was used, the measured temperature fields clearly incorporate heat conduction effects. Due to this, the surface temperatures were visibly influenced around the hole exits which results into non-adiabatic wall conditions. Furthermore, the use of the plenum temperature to calculate film effectiveness neglects the heat loss through the cooling hole walls.

A suitable approach to assess effects driven by solid heat conduction is the use of a FE-thermal analysis (see chapter 6). As a final result from the simulations, the computed heat conduction pattern on the solid wall surface can be applied to correct the temperature measurements. The coolant temperature at the hole exit is evaluated by a heat flux balance across the hole walls.

With that the adiabatic film effectiveness for an inverse coolant-to-main flow temperature ratio is computed using equation (2.7). Herein T_{aw} is the wall temperature after correction with the FE-results and T_c is the coolant discharge temperature accounting for the estimated hole heat loss.

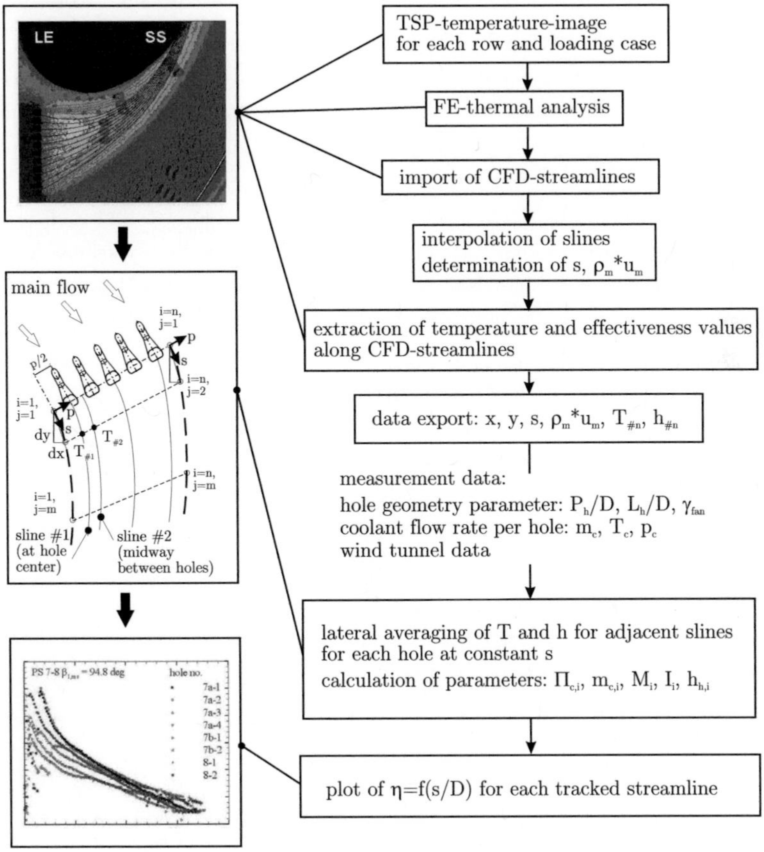

Figure 5.2: Data reduction methodology for the evaluation of streamwise film effectiveness using CFD-calculated surface streamlines

After correction, the film effectiveness distributions are transformed into the airfoil coordinate system for import and extraction of the streamwise data along CFD-calculated surface streamlines. Its origin is located close to the corner formed by the blunt trailing edge and the pressure side of the airfoil (see also Fig. 3.5).

Each of the derived data sets for a film row test is plotted versus the normalized streamwise distance s/D. The local hole blowing rate for a specific row is calculated using the main flow mass velocity at each hole injection location $(\rho_m \cdot u_m)_i$ and the coolant mass flow rate per hole. To assess the streamwise evaluation of the film effectiveness, the following methodology, as displayed schematically in Fig. 5.2, is applied to each investigated film row and for all loading cases. Due to the different endwall locations of the film rows and strong enhancement of the near-wall flow field with increasing incidence, each row test case has to be treated separately.

The initial point for the data reduction process is the wall temperature field for a specific film row test case. At first, the global coordinate system of the measurement data and the CFD-solution are synchronized, so that the surface streamlines can be imported. Two different streamlines for each hole are provided by the CFD-computations with one initiating at the hole center and the other one midway between two holes. Two additional streamlines are added spaced half pitch from the hole center of the first and last hole of a film-cooling row. The streamlines are then tracked along their curvature following the local streamwise coordinate system $s-p$. The origin of this coordinate system is defined at the location where the hole axis emerges through the endwall. The s-axis is orientated tangentially to the streamline direction.

The extracted surface temperatures and film effectiveness are exported to be combined with flow field variables computed by the simulations like the main flow pressure, density and velocity. The streamlines are interpolated at constant streamwise distances Δs. The temperatures between the streamlines at the hole center and midway between adjacent holes are laterally averaged. Combined with auxiliary wind tunnel data from the row tests, as e.g. the individual hole coolant flow rate, plenum pressure p_{pl}, plenum temperature T_{pl} and the film hole geometry, the hole blowing rates and dimensionless streamwise distance s/D are calculated. In order to use streamlines for the predefined TSP-data analysis that correctly indicate the near-wall flow direction even at enhanced loading conditions, detailed numerical simulations of the cascade flow field are required. The corresponding numerical model is therefore explained throughout the following section.

5.3.1 Numerical Model

For the computation of the cascade flow field applying numerical methods, a closed system of equations must be assembled which mathematically describes the occurring physical processes. Here, the governing equations for the conservation of mass, momentum and energy are used which are solved in the discretized computational domain in space and time.

Subsequently, the conservation laws are presented during the following section in its differential form and in the notation of cartesian coordinates. It is further followed by the description of the numerical approaches on turbulence modeling whose are required to close the Reynolds-Averaged Navier-Stokes-Equations (RANS). Only the used two-equation model is explained while other forms of turbulence modeling as zero, one equation models or Reynolds-Stress turbulence models were not considered. Furthermore, detailed informations are given on the computational domain and the mesh which are used for the presented work and selected numerical results are compared to available experimental data.

Continuity equation. The continuity equation describes the conservation of mass in a system. It defines that the mass of an infinitesimal fluid element is constant while it moves through an arbitrary chosen control volume fixed in space. The differential form of the continuity equation is given by

$$\frac{\partial \rho}{\partial t} + \nabla \cdot (\rho \mathbf{u}) = \frac{\partial \rho}{\partial t} + \frac{\partial \rho u_i}{\partial x_i} = 0. \tag{5.36}$$

Momentum equation. The momentum is a vector and is defined in general as product of velocity and the total mass of an element. The total momentum being described as vector sum of all initial momentums is constant in a closed inertial system. Due to that, the conservation law of the momentum states that the rate of momentum change for a material fluid element with respect to time equals to the sum of all driving forces. Applying Newton's second law as a fundamental physical principle to a flow model leads to

$$\frac{d\mathbf{I}}{dt} = \frac{d}{dt} \int \rho \mathbf{u} dV = \sum \mathbf{F_s} \tag{5.37}$$

where the sum of the driving forces $\sum \mathbf{F_s}$ can be divided into surface and body forces. Surface forces directly act on the surfaces of the fluid element. They represent the effect of the surrounding flow conditions and either result from the pressure distribution imposed by the outside flow or from the shear- and normal-stress distribution which act on the elemental surface by means of friction. Furthermore, body forces act on the volumetric mass of the fluid and are induced by force fields as gravitation, magnetism or electrical fields. Following the derivation proposed by Wendt [136], the differential form of the momentum equation is given by

$$\frac{D\rho\mathbf{u}}{Dt} = \frac{\partial (\rho \mathbf{u})}{\partial t} + \nabla \cdot (\rho \mathbf{u} \times \mathbf{u}) = -\nabla p + \nabla \cdot \tau + \rho \mathbf{f_s} \tag{5.38}$$

and in tensor notation, respectively

$$\frac{\partial \rho u_i}{\partial t} + \frac{\partial \rho u_i u_j}{\partial x_j} = -\frac{\partial p}{\partial x_i} + \frac{\partial \tau_{ij}}{\partial x_j} + \rho f_{s,i}. \tag{5.39}$$

The surface forces are composed of the pressure forces $\frac{\partial p}{\partial x_i}$ and the shear stress $\frac{\partial \tau_{ij}}{\partial x_j}$. The variable f_i describes the body forces. According to Newton, the stress tensor τ_{ij} is proportional to the strain rate denoting such fluids as Newtonian fluids. Furthermore, the fluid properties are assumed to be isotropic with no preferred direction.

Following Stokes hypothesis that the bulk viscosity coefficient μ' must diminish to achieve identity between the thermodynamic pressure and the mean mechanical pressure in a deforming viscous fluid, gives $\mu' = (\psi + 2/3\mu) = 0$ with the dynamic viscosity induced by shear μ and dilatational viscosity ψ. Physically, this factor is connected to the dissipation mechanism during a volume change at a finite rate. Taking into account these explanations, the tensor for the shear stress of a Newtonian, isotropic fluid can be written as

$$\tau_{ij} = \mu \left(\frac{\partial u_i}{\partial x_j} + \frac{\partial u_j}{\partial x_i} - \frac{2}{3}\delta_{ij}\frac{\partial u_k}{\partial x_k} \right) \tag{5.40}$$

with δ_{ij} the Kronecker identity tensor which equals $\delta_{ij} = 1$ for $i = j$ and $\delta_{ij} = 0$ for $i \neq j$. The dynamic viscosity coefficient is a physical property for any used fluid and can be extracted from experimentally obtained data sets. In this case, the vector for the body forces f_i is limited to the gravitation field of the earth. Its influence on the present flow is therefore neglected. Substituting (5.40) in (5.39), the momentum equations for a Newtonian, isotropic fluid, also denoted as Navier-Stokes-equations, can be derived

$$\frac{D\rho u_i}{Dt} = \rho f_{s,i} - \frac{\partial p}{\partial x_i} + \frac{\partial}{\partial x_j}\left[\mu \left(\frac{\partial u_i}{\partial x_j} + \frac{\partial u_j}{\partial x_i} - \frac{2}{3}\delta_{ij}\frac{\partial u_k}{\partial x_k} \right) \right]. \tag{5.41}$$

Energy equation. Based on the first law of thermodynamics and the conservation of energy, the change of energy inside a infinitesimally small moving fluid element over a time increment dt equals the sum of the heat flows in or out of the element and the rate of work done on the element caused by body and surface forces. The heat flows of an element are due to volumetric heating such as absorption or emission of radiation and heat transfer across the element surfaces due to the temperature gradients. The sum of the individual heat flows can be derived by integrating the heat fluxes over the entire element surface. Based on Fourier's law,

$$\dot{q}_i = -\lambda \nabla T = -\lambda \frac{\partial T}{\partial x_i} \tag{5.42}$$

the energy equation can be derived in its conservative form as

$$\frac{\partial \rho E}{\partial t} + \nabla \cdot (\rho E \cdot \mathbf{u}) = \rho \dot{q} + \nabla \cdot (\lambda \cdot \nabla T) + \nabla \cdot (\mathbf{u} \cdot \tau) - \nabla \cdot (\mathbf{u} \cdot p) + \rho \mathbf{f_s} \cdot \mathbf{u} \tag{5.43}$$

and in tensor notation, respectively

$$\frac{\partial(\rho E)}{\partial t} + \frac{\partial(\rho u_i E)}{\partial x_i} = \rho\dot{q} + \frac{\partial}{\partial x_i}\left[\lambda\frac{\partial T}{\partial x_i}\right] + \frac{\partial u_j \tau_{ij}}{\partial x_i} - \frac{\partial p u_i}{\partial x_i} + \rho f_{s,i} u_i. \tag{5.44}$$

The terms on the equation's right hand side describe the heat flow due to volumetric heating $\rho\dot{q}$, the heat transfer due to temperature gradients $\nabla \cdot (\lambda \cdot \nabla T)$ and the work done by the shear stresses and friction $\nabla \cdot (\mathbf{u} \cdot \tau)$, by pressure on the element surfaces $\nabla \cdot (\mathbf{u} \cdot p)$ as well as by body forces $\rho \mathbf{f_s} \cdot \mathbf{u}$.

The preceding equations apply to a viscous flow that includes dissipative transport phenomena of viscosity and thermal conduction. Mass diffusion is not considered in this case since the used fluid is limited to a homogenous, non-chemically reacting gas. They form a coupled system of non-linear partial differential equations which contains five equations for the six unknown variables ρ, p, u_i and E. Therefore, the transport equations must be closed by a constitutive equation for the density. Assuming an ideal gas behavior, the density is calculated with the ideal gas law and the specific heat capacity is a function of temperature. The equation of state is then

$$p = \rho R T = (\kappa - 1)\rho e \tag{5.45}$$

with R the specific gas constant and p the static pressure. The temperature T is an additional unknown variable which is derived from a thermodynamic relation between state variables. The definition of the total energy and the internal energy e for an ideal gas is given by

$$E = \rho\left(e + \frac{1}{2}\mathbf{u}^2 + gh\right), \tag{5.46}$$

$$e = \frac{p}{\rho}\left(\frac{1}{\kappa - 1}\right). \tag{5.47}$$

This finally closes the mathematical problem by forming a system of seven equations with seven unknown variables that can be solved numerically.

Numerical modeling of turbulence. In a turbulent flow field, the describing flow quantities such as velocity components, pressure and temperature are exposed to permanent fluctuations. These quantities are of three-dimensional and highly unsteady character and will strongly vary in their spatial extent. As an approach to describe such fluctuations, O. Reynolds in 1894 [103] introduced the concept of statistical turbulence analysis. Based on that, an instantaneous flow quantity is decomposed into its time-averaged part $\bar{u}$ and its fluctuating part u'.

$$u_i(\mathbf{x}, t) = \overline{u_i}(\mathbf{x}, t) + u_i'(\mathbf{x}, t). \tag{5.48}$$

The definition of the time-averaged quantity for a statistically time-invariant (steady) flow is given by

$$\overline{u_i}(\mathbf{x}, t) = \lim_{T\to\infty} \frac{1}{T} \int_0^T u_i(\mathbf{x}, t)dt. \tag{5.49}$$

The decomposition is here shown for the velocity but is applied to all flow quantities, accordingly. Based on the definition given above, it is required that the time scales of the fluctuations are significantly smaller than the time scales of its mean quantities. This results into a sharp separation of the frequency spectra between the mean flow field and its fluctuating components. If the latter is not fulfilled, this leads to a spectral-gap-problem [137]. Applying the averaging approach to the Navier-Stokes-equation (5.40) and the continuity equation (5.37), one obtains the Reynolds-averaged Navier-Stokes-equations (RANS) as given by

$$\frac{\partial \rho \bar{u}_i}{\partial t} + \frac{\partial \rho \bar{u}_i \bar{u}_j}{\partial x_j} = -\frac{\partial \bar{p}}{\partial x_i} + \mu \frac{\partial^2 \bar{u}_i}{\partial x_j^2} - \frac{\partial \rho \overline{u_i^{'} u_j^{'}}}{\partial x_j} \tag{5.50}$$

All linear fluctuation terms disappear while the quadratic non-linear term $\overline{u_i^{'} u_j^{'}}$ that is related to the non-linear convection term remains in the equation. These non-linear terms of the fluctuating quantities are denoted as turbulent or Reynolds stresses which are comparable to the molecular stresses induced by friction. To achieve closure of this system of governing equations, further relationships are required to express the still unknown Reynolds stress tensor $\tau_{ij}^R = \rho \overline{u_i^{'} u_j^{'}}$.

An approach to solve this problem was found by Boussinesq [14] who introduced the concept of turbulent eddy viscosity. This eddy viscosity hypothesis proposes that the Reynolds stress tensor can be related to the mean velocity gradient where the newly introduced turbulent eddy viscosity μ_t is used as a proportionality factor.

$$-\overline{u_i^{'} u_j^{'}} = \mu_t \left[\frac{\partial \overline{u_i}}{\partial x_j} + \frac{\partial \overline{u_j}}{\partial x_i} \right] - \frac{2}{3} k \delta_{ij}. \tag{5.51}$$

Here μ_t has the same dimension as the molecular viscosity but is not a physical property of the fluid. It depends instead on the local turbulence structure. Furthermore the term $k = 1/2\overline{u_i^{'2}}$ named turbulent kinetic energy is introduced which represents the principal diagonal of the Reynolds stress tensor. This approach introduces the eddy viscosity as new unknown variable to replace the quantities of the Reynolds stress tensor. The eddy viscosity parameter is assumed to be decoupled from the mean velocity field but explicitly depends on characteristic values that describe the turbulent flow structure.It follows that the eddy viscosity can be related to a characteristic length L_t and velocity scale u_t of the turbulent movement.

Turbulence models using this approach would be commonly known as eddy viscosity models. Different sub-models exist which can be distinguished by their number of solved equations. Considering algebraic or zero-equation-models like the mixing-length model, no transport equation is solved. Instead, it directly relates the Reynolds stresses to the mean velocity field. One-equation models like the Spalart-Allmaras-model is a more basic model solving only a single transport equation for the turbulent viscosity. Sophisticated two-equation models as the $k-\epsilon$- and $k-\omega$ model incorporate several differential equations to solve for the production of turbulent kinetic energy k, turbulent eddy dissipation ϵ and turbulent frequency ω, respectively.

Considering the $k-\omega$ model of Wilcox [138], its definition assumes that the turbulent viscosity is linked to these quantities using the following relation

$$\mu_t = \rho \frac{k}{\omega}. \tag{5.52}$$

The model solves two separate transport equations for the turbulent kinetic energy k and the turbulent frequency ω

$$\frac{\partial k}{\partial t} + \frac{\partial k \bar{u}_i}{\partial x_i} = \frac{\partial}{\partial x_i}\left[\left(\mu + \frac{\mu_t}{\sigma_k}\right)\frac{\partial k}{\partial x_i}\right] + \frac{P_k}{\rho} - \beta' k\omega \tag{5.53}$$

$$\frac{\partial \omega}{\partial t} + \frac{\partial \omega \bar{u}_i}{\partial x_i} = \frac{\partial}{\partial x_i}\left[\left(\mu + \frac{\mu_t}{\sigma_\omega}\right)\frac{\partial \omega}{\partial x_i}\right] + \alpha \frac{\omega}{k}\frac{P_k}{\rho} - \beta\omega^2 \tag{5.54}$$

with $\alpha = 5/9$, $\beta = 0.075$, $\beta' = 0.09$, $\sigma_k = 2$ and $\sigma_\omega = 2$ as heuristic model constants. P_k is the production rate of turbulence due to viscous forces and is defined as

$$P_k = \mu_t \nabla \mathbf{u} \cdot (\nabla \mathbf{u} + \nabla \mathbf{u}^T) - \frac{2}{3}\nabla \cdot \mathbf{u}(3\mu_t \nabla \cdot \mathbf{u} + \rho k). \tag{5.55}$$

A more accurate calculation of the Reynolds stress tensor can be achieved applying Reynolds stress turbulence models. Those models must not rely on the eddy viscosity hypothesis solving transport equations for all components of the Reynolds stress tensor and the dissipation rate. By modeling all stress components using either algebraic or differential equations, this approach is more suitable for computing highly complex and anisotropic flows. Although this approach provides increased numerical accuracy of the obtained results, the computational costs by means of increased run time and occurring convergence instabilities are certainly disadvantageous. Therefore, the more robust $k-\omega$ turbulence model is used for the presented numerical work.

5.3.2 Computational Domain

Numerical simulations were performed with the commercial flow solver ANSYS-CFX™ 5.0 for the model geometry of the nozzle guide vane used during the endwall cooling experiments. The flow solver employs a conservative finite element-based control volume method. The spatial discretization scheme is of second order and is used for all transport equations except the turbulence quantities. Steady-state numerical calculations are performed for all test cases using single component formulation with air as fluid. The used turbulence model is the low-Reynolds formulation of the two equation $k-\omega$ turbulence model.

The purpose of those numerical experiments were twofold. On the one hand side, it provides informations on the near-wall flow field in form of surface streamlines that could be used to determine streamwise film effectiveness as function of loading. The simulations further give detailed insight to the flow behavior in regions that are not accessible during the experiments.

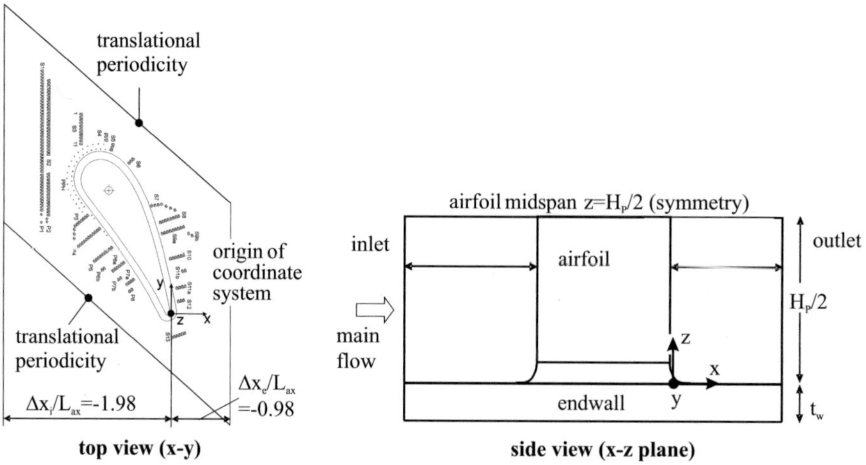

Figure 5.3: Computational domain of the vane cascade

On the other hand, the CFD-model can be used to estimate endwall heat transfer rates which can be incorporated as boundary conditions into a FE-model. This approach is used to estimate heat conduction effects on the measured wall temperature fields (see also chapter 6). The computational domain used for the numerical investigations is shown in Fig. 5.3.

For the numerical simulations, the pitchwise extent of the computational domain resolves a periodical section of the airfoil within one pitch. Due to symmetry considerations, only half of the profile extending from the endwall up to midspan was modeled to reduce the size of the computational domain. The location of the inlet and outlet plane was selected to $\Delta x/L_{ax}$ = -1.98 (inlet) and $\Delta x/L_{ax}$ = 0.98 (exit). The fluid domain extends one axial chord length upstream of the leading edge and one axial chord length downstream of the trailing edge. A symmetry plane in spanwise direction was defined at airfoil midspan z_{ms}. The fillet geometry with its different radii at the vane junction was also implemented into the CFD model.

Following constraints are required for the numerical mesh. It should be capable of resolving the secondary flow pattern near the endwall, so the effect of loading on the film-cooling injection is accurately predicted. The computational grid must further provide a detailed mesh resolution of the flow around the airfoil profile to resolve the wall boundary layers. Moreover the mesh size should be adapted in the free-stream region (outside of the boundary layer) to reduce the computational and the meshing efforts. Last but not least, the low-Reynolds formulation for the RANS equations are to be applied without the need to use wall laws.

The geometry was meshed using a hybrid approach by choosing unstructured tetrahedral elements for the free-stream region and structured hexahedral elements for the wall regions of the profile and endwall boundary layer. The computational domain is divided into three major regions: 1) the main flow field, 2) the region of the platform wall boundary layer and 3) wall region around the airfoil profile. The computational mesh was created using ANSYS ICEM

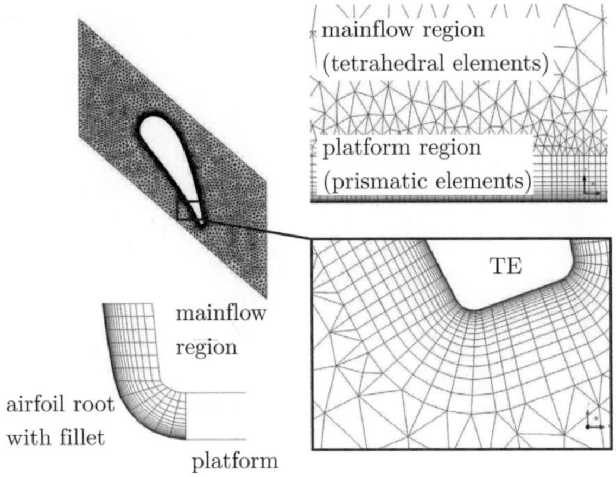

Figure 5.4: Mesh structure of computational grid

CFD^{TM}10.0. The grid size of the first mesh node of this model is 1*10^{-5}m. This leads to a normalized wall distance z^+ in the order of 1 for the profile boundary layer and is determined to the order of 1-3 for the platform region. The expansion ratio of the cell width is set to 1.25 for the direction perpendicular to the endwall. For the connection of the hexahedral wall mesh to the mesh of the main flow region, a grid layer of pyramidal cells was implemented along the mesh interface. Three different grids were created to study grid independency of the computational model with different grid size. They are referred to 1) a coarse grid with a total node count of 1.8 Mio. nodes and 2) medium grid with 3.45 Mio. nodes and 3) a fine mesh with approx. 9 Mio. nodes. Table 5.1 gives a summary of the grid size per mesh region for all three different meshes. The majority of the cells of about 50 % of the total grid size is put in the mesh for the platform region with the film holes in order to resolve the secondary flow pattern.

Dirichlet boundary conditions were applied to the inlet plane in terms of the measured pitchwise-averaged distribution of the total pressure, the inlet flow angle and the free-stream turbulence intensity for all three cascade angle configurations. A constant total temperature is applied to the inlet boundary interface. At the outlet plane, a constant static pressure condition of $p_{amb} = 101325\,Pa$ is imposed. Due to the large aspect ratio of the cylindrical airfoil, the endwall loss regions do not merge at midspan. Therefore a symmetric flow is assumed and the computational domain was reduced to a half-span simulation between the platform and the symmetry plane located at airfoil midspan. The solid-wall boundaries of the platform and airfoil profile are modeled as impermeable and adiabatic. At all walls, a no slip flow condition

mesh region / resolution	coarse	medium	fine
	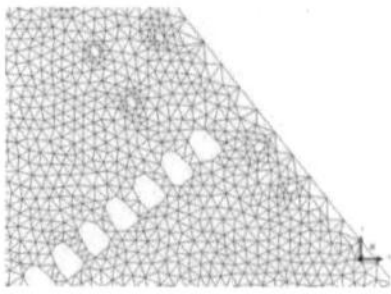	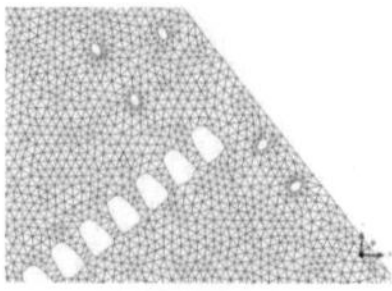	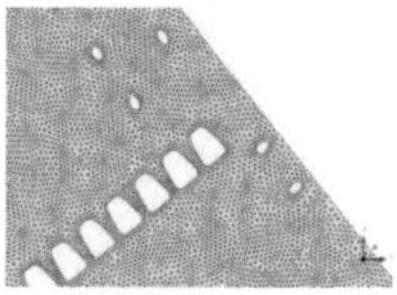
main flow region	0.35 Mio.	0.75 Mio.	1.50 Mio.
profile region	0.45 Mio.	0,95 Mio.	0.95 Mio.
endwall region	1.0 Mio.	1.75 Mio.	ca. 6.5 Mio.
total sum	1.8 Mio.	3.45 Mio.	approx. 9 Mio.

Table 5.1: Number of grid nods per mesh region for the computational grids at different mesh size

is applied. Translational periodic conditions are employed on the pitchwise interfaces with a 1:1 mesh connectivity. For the calculation of the endwall heat transfer rates, a Dirichlet boundary condition was applied in terms of a constant temperature difference.

The fluid domain is modeled as a variable composition mixture of two types of gases as e.g. air and carbon dioxide but also other gases with differing physical properties could be implemented. The mass fraction can be adjusted from 0 % to 100 % for each component in accordance to the desired density ratio. The ideal gas equation of state is used to describe the gas behavior. The dynamic viscosity μ for air is evaluated as function of the gas temperature and pressure based on a formulation proposed by Lemmon and Jacobson [78]. Significant results of the numerical simulations and their comparison to the obtained experimental data are shown in chapter 8.1.7.

Prior to the detailed CFD calculations, the concept of the grid convergence index (GCI) was used to examine the ordered discretization error based on the theory of a generalized Richardson extrapolation. This method was first proposed by Roache et al. [104–106]. The GCI is computed by three levels of grid refinement. A minimum of grid refinement ratio of r > 1.1 should be maintained to differentiate the discretization error from other error sources. This method obtains an estimate of the continuum value for a computed quantity f at zero grid spacing while the simulation is assumed to be in the asymptotical range of convergence. Therefore, a simulation quantity is expressed by a series expansion applied to two solutions with different spacing h_1 and h_2 with h_1 being the finer spacing.

In this case, the transport equations for mass, momentum and energy were discretized in space using a second order accurate scheme. On the contrary, the transport equations for the turbulence quantities were only first order accurate. The GCI is a measure of %age how far away the actual solution with a specific spacing is from the asymptotical value at zero spacing. The methodology can be applied to either all grid points or to functional variables of the numerical solution. The GCI for the fine grid is defined as

$$GCI_{12} = \frac{S_f \, |\epsilon|}{(r^p - 1)} \tag{5.56}$$

with r the grid refinement ratio, p the order of convergence and S_f a safety factor which is recommended to be 1.25 for the comparison of three grids. Further ϵ describes the relative error between two solutions

$$\epsilon = \frac{f_2 - f_1}{(f_1)}. \tag{5.57}$$

For the evaluation of the GCI, the flow field is computed on three different grids with refinement ratios of approximately 3. Table 5.2 summarizes the results of the grid convergence study in terms of the area-averaged total pressure loss coefficient ζ_2 and the profile loss coefficient ζ_p.

Grid	r	p	ζ_2	$\zeta_{2,h=0}$	GCI	p	ζ_p	$\zeta_{p,h=0}$	GCI
1	1	1.73	0.068842		0.00031	1.72	0.051110		0.00055
2	3	1.73	0.068941	0.068825	0.00211	1.72	0.051239	0.051088	0.00375
3	6	1.73	0.069609			1.72	0.052120		

Table 5.2: Results of the grid convergence analysis

The grid convergence index for the area-averaged loss coefficient with respect to the comparison of the grids 1-2 is found to be quite small at 0.031 %. Due to slightly lower order of convergence, the GCI_{12} for the profile loss is not remarkably higher with 0.055 %. Considering the coarser grids 2-3, the GCI_{23} increases to 0.211 % for total pressure loss coefficient and to 0.375 % for the profile loss. Using following relationship

$$GCI_{23} = r^p \cdot GCI_{12} \tag{5.58}$$

two grid levels can be compared with respect to the order of convergence. If it is nearly unity, it can be verified that the computational solutions on each grid level lie within the asymptotic range. Applying this approach to the present results yields a value of 0.99855 for the total loss coefficient and 0.99747 for the profile loss. Hence, the solutions are well in the asymptotic range of convergence.

6 Thermal Analysis using FEM

The application of main flow-coolant temperature gradients to determine film effectiveness distributions certainly imply heat conduction and radiation effects as long as the investigated wall is not perfectly adiabatic. Due to small inverse temperature ratios ($T_R < 1.12$, heated coolant / cold main flow) applied in the present film-cooling experiments, influences from heat radiation can be considered to be small. Therefore, they will be neglected for later discussions. To minimize the conduction effects, the test plates are manufactured from low conductivity material. Although the heat conduction is drastically reduced by using such materials, the surface-temperature measurements still showed influence of the three-dimensional heat conduction through the test plate.

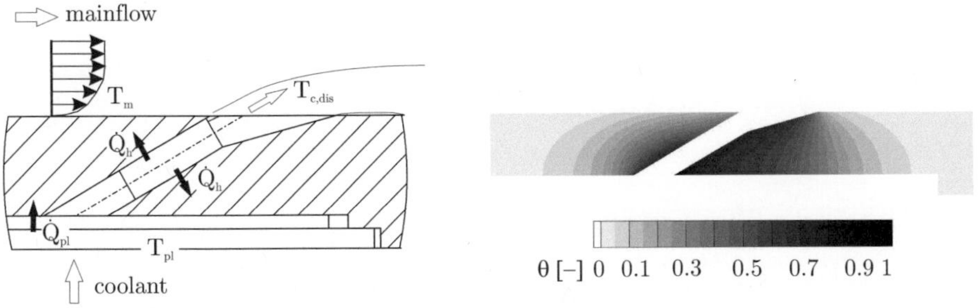

Figure 6.1: Heat flow situation in a test plate with finite thermal conductivity (left) and non-dimensional temperature distribution due to heat conduction (right)

A schematic of the thermal situation in a film-cooled wall is presented in Fig. 6.1. The film effectiveness, which is derived from the raw surface temperature measurements, is defined for compressible flows according to equation (2.6). Due to the actual low-speed flow conditions in the wind tunnel, the recovery temperature is replaced by the static inlet temperature of the main flow. The measured wall temperature T_w is affected by the heat transfer induced by the injected coolant and the endwall heat conduction. Especially around the hole exits, the temperature distribution can tremendously be changed due to the reduced wall thickness.

Since an adiabatic test surface is needed to separately measure film effectiveness from the heat transfer following the superposition approach, it is required to eliminate the effects of heat conduction by an appropriate thermal analysis. Assuming the heat conduction onto and inside the test plate shall be zero, the film-cooling effectiveness is replaced by the adiabatic film-cooling effectiveness given by equation (2.7).

The adiabatic film effectiveness would differ due to conduction and possibly occurring radiation effects in high-temperature test-rigs. For the correct expression of the adiabatic film effectiveness the coolant temperature at the hole discharge should be used. Using the plenum temperature neglect the heat transfer inside the cooling holes. For the applied inverse temperature ratio, the exit temperature of the coolant is reduced due to heat losses across the hole and plenum walls. Nevertheless, measuring the discharge temperature without disturbing the jet flow field is complicated. Hence, the heat loss is estimated by a simplified heat balance.

To assess the effects of heat conduction on the wall temperatures, a steady-state thermal analysis is carried out using the Finite-Element method. Based on results of the computations, a correction method is applied to the measured temperature images to eliminate these conduction effects. The FE-based thermal analysis is focused to capture the conductive wall heat fluxes directly upstream of the holes which primarily influences the wall temperature fields. It is further used to estimate heat loss across the plenum and the cooling hole walls calculating the coolant exit temperature. This procedure is applied to all experimental film-cooling data, either coming from the flat-plate studies or the endwall film-cooling investigations.

6.1 FE-Models

6.1.1 Flat Plate Film-Cooling

Hence, a detailed FE-model was created enclosing all holes of the film row to represent the actual conditions in the flat-plate film-cooling tests. Using this model, a detailed thermal analysis was performed for each blowing rate and density ratio using the commercial Finite-Element code ABAQUS^{TM}. A schematic view on the FE-models for both cooling-hole configurations is displayed in Fig. 6.2 along with a close-up on the created FE-mesh inside and at the hole exit.

The axial extend of the model is 150 mm upstream and downstream of the hole exits. This reduces the influence of the interface temperature by conduction effects. With that these boundaries can be employed with an adiabatic wall condition. The lateral extent is set according to test duct's width of 160 mm. For the mesh regions around the holes and the plenum chamber, where large temperature gradients are expected, the mesh was refined to accurately resolve the wall heat fluxes. Also, the mesh resolution on the test surface was adopted to be close to the spatial resolution of the CCD-camera of 9 $pixel/mm$. This results in a total element count of 472.155 tetrahedral elements for the cylindrical-hole mesh. For the test case with fan-shaped holes, the mesh contains 325.683 tetrahedrons. The hole sections are resolved with 3640 triangular elements resulting in a minimal edge length of 0.23 mm. A similar resolution is achieved for the model with the cylindrical holes containing 2188 triangular elements per hole

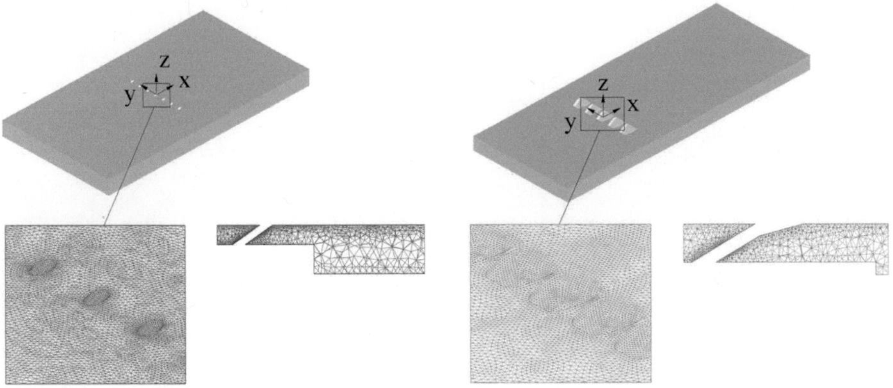

(a) cylindrical hole geometry (b) laid-back fan-shaped hole geometry

Figure 6.2: Schematic of flat plate Finite-Element model

with a minimal edge length of $0.22\,mm$.

The hole coordinate system is identical to the definition used for the experimental data and is displayed in Fig. 6.2. The origin is settled at the intersection point of the hole axis with the surface for each row's center hole. The x-axis is oriented in streamwise direction and the y-axis in pitchwise direction. Adiabatic conditions are imposed onto front and side faces of the mesh. For the element's faces inside the film holes, on the test wall and the plenum, a constant heat transfer coefficient is applied which is estimated from correlations (see also section 6.2.2 for details).

6.1.2 Endwall Film-Cooling

To obtain the conductive heat-flux pattern on the cascade endwall, a detailed FE-model of the entire test plate with all film holes was created. The simulations were performed also using the commercial Finite-Element code ABAQUS^{TM}. The geometry of the test plate was extracted from the CAD-model and was imported into the meshing tool. The test plate with its numerous film holes was then meshed with the grid generator ICEM^{TM} to create an unstructured mesh with tetrahedrons of a total node count of 527.938 nodes. This yields a total count of 2.88 Mio. tetrahedral elements due to the element's high stacking density. The regions of the hole exits are refined in order to resolve the heat flux pattern in detail. Each film hole exit contains between 345 and 440 triangular surface elements which results into a range of 197 to 246 nodes per hole. To display the element quality for the film holes, the minimum length of the smallest element side and the minimum internal angle of the trias has been determined. The minimum edge

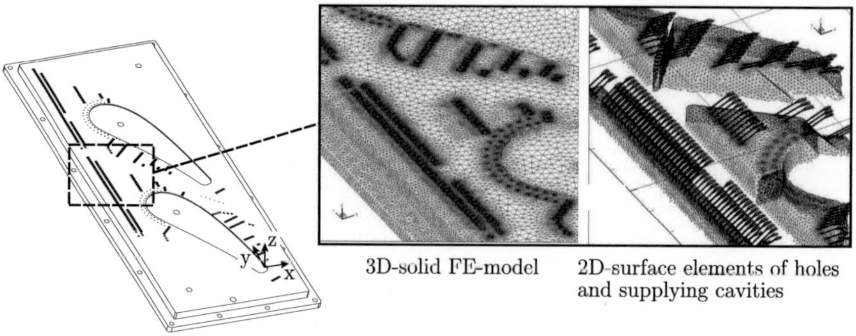

Figure 6.3: Schematic of endwall Finite-Element model

length varies between 0.12 mm and 0.22 mm and the angle does not exceed 18°. The maximum aspect ratio of the elements is not larger then 3.3. An ideally-shaped triangular element would a value of 1.0 for this ratio between the element's shortest and longest side. After the meshing process, the model was exported in a Nastran-format to be importable into Hypermesh^{TM}. This tool was used to generate the proper input-files for the FEA-solver ABAQUS^{TM}. Furthermore, individual node and element sets for the cylindrical and diffuser hole sections were created with Hypermesh^{TM} in order to apply different boundary conditions.

6.2 Boundary Conditions

6.2.1 Flat Plate

Due to a constant main flow velocity during the flat-plate experiments, a constant heat transfer coefficient is applied for all wall surface elements. This value is derived from the heat transfer correlation for a turbulent flow over a flat plate as given in Kays et al. [67]

$$h_m = 0.0287 \left(c_{p,m} \rho_m u_m\right) Re_m^{-0.2} Pr_m^{-0.4}. \tag{6.1}$$

The characteristic length for the main flow Reynolds-number

$$Re_m = \frac{u_m \Delta x_h}{\nu_m} \tag{6.2}$$

is the distance from the duct inlet to the hole exits which equals $\Delta x_h = 0.53$m. The main flow velocity

$$u_m = \sqrt{\frac{2\left(p_{tm} - p_m\right)}{\rho_m}} \tag{6.3}$$

and the main flow density

$$\rho_m = \frac{p_m}{T_m} R_m \tag{6.4}$$

is obtained from the measured quantities for the main flow static pressure p_m, temperature T_m and velocity head $p_{tm} - p_m$ during the test runs. The heat transfer solution in equation (6.1) for a constant free-stream velocity and surface temperature of a turbulent external boundary layer is valid for Prandtl numbers between 0.5 and 1.0 and Reynolds numbers between $5{\cdot}10^5$ and $5{\cdot}10^6$. For all experimental cases, the main-flow Reynolds number is within that threshold and equals $Re_m = 7.94{\cdot}10^5$.

The Dittus-Boelter-relationship for fully-developed turbulent flows inside a pipe was used to estimate the heat transfer coefficient for the flow inside the holes. It is given in Kays et al. [67] for tube length-to-diameter ratios of $10 < L_h/D < \infty$

$$h_h = \frac{0.023 Re_{c,D}^{0.8}\,\mathrm{Pr}_c^{0.4}\,\lambda_c}{D}. \tag{6.5}$$

This equation is valid for Reynolds numbers Re_D based on the hole diameter between 10^4 and 10^5 and Prandtl numbers of 0.6 ... 6. In the experimental cases the Reynolds number for the cylindrical holes ranges between 3000 to 10000. For the diffuser holes, the Reynolds number is slightly higher due to the larger hole diameter and increases with blowing rate from 4000 to 18000. To consider the enhanced heat transfer at the hole entrance, a constant correction factor of 1.8 is applied to the heat transfer coefficient estimated with the Dittus-Boelter-equation. Thus, the heat transfer coefficient inside the cylindrical hole section is obtained by

$$h_h = 1.8\frac{0.023 Re_{c,D}^{0.8}\,\mathrm{Pr}_c^{0.4}\,\lambda_c}{D}. \tag{6.6}$$

As the coolant flow velocity is decreased inside the diffuser section recovering static pressure, the heat transfer is also reduced. To consider this behavior in the simulations, the coolant flow velocity is estimated by the area ratio between the exit of each diffuser part and the cross-section of the metering part of the hole. For the lateral diffuser section, this ratio is determined at its intersection with the axial diffusion part at a hole length of $L_h/D = 5$. The exit area of the axial diffuser is defined at the imaginary intersection of the hole with a plane which is perpendicular to the hole axis. This plane is located at the point where the hole axis emerges through the surface at a hole length of $L_h/D = 6$. This concept of the diffuser area ratio was introduced by Goldstein et al. [42] in 1974 for more convenient comparison of different fan-shaped geometries.

Considering real flow conditions, the coolant flow is expanded continuously inside the diffuser. This continuous decrease in flow velocity gives reduced turbulence along the opening and leads to a steady decrease in heat transfer. For the cases considered here, this effect is simplified to be constant for each shaped hole section. Herein the maximum opening area ratio is used which slightly overestimates the effect in the lower diffuser parts. It is calculated to $A_R = 2.9$ for the lateral diffuser part. The axial part exhibits a larger area ratio of 4.7. The reduction factor that is applied to the heat transfer coefficient for each diffuser is then 0.43 for the lateral opening and 0.29 for the axial opening.

6.2.2 Cascade Endwall

Performing a thermal analysis calculating the conductive wall heat fluxes on a cooled cascade endwall requires the application of proper boundary conditions. Moreover, they should reflect the actual measurement conditions and therefore the driving potentials for the heat conduction. Heat transfer rates on the hot gas path side change significantly due to the fluid's acceleration in the airfoil passage. Furthermore, they receive visible enhancement due to the near-wall secondary flow structure which may not be estimated by heat transfer correlations. Out of these reasons, a sophisticated numerical approach was chosen simulating the gas path heat transfer coefficients on the endwall surface by the existing CFD-model. The latter is then applied as boundary condition on the endwall in the FE-model.

The effect of heat transfer enhancement caused by the presence of film-cooling injection, which can be of the order of 10 to 20 %, is hereby neglected. The computational grid includes only the hole exits on the endwall without discrete modeling of the entire flow path through the holes. This is done in order to significantly reduce the meshing efforts, the total element count and overall computational costs. Note that since the FE-based correction method focuses on the upstream region of the film holes, the effect of enhanced heat transfer induced by the film jets still remains in the measured wall temperature fields. The spatial heat transfer coefficients have been derived from steady-state numerical computations using ANSYS CFX^{TM} as commercial CFD-tool. A detailed description of the numerical model and its boundary conditions can be found in section 5.3.1.

Two separate computational runs are performed. In the first one, adiabatic wall conditions are applied which delivers the adiabatic wall temperature distribution across the endwall. The followed run is conducted by imposing a Dirichlet boundary condition on the endwall surface. Here, a constant temperature difference is set between the walls and the flow region. This simulation predicts the resulting wall heat flux which is used to calculate the heat transfer coefficients for a turbulent, accelerated flow through an cascade as given by

$$h_m = \frac{\dot{q}_w}{(T_w - T_{aw})}. \tag{6.7}$$

The CFD-results of the heat transfer coefficient are furthermore compared to the correlation for a turbulent flow over a flat plate which is given by Kays et al. [67]. It is defined as

$$h_{m,corr} = 0.0287\,(c_{p,m}\rho_m u_{m,x})\,Re_{m,x}^{-0.2}Pr_m^{-0.4} \tag{6.8}$$

with

$$Re_{m,x} = (u_{m,x}L_p\rho_m)\,/\mu_m \tag{6.9}$$

where u_x is the spatial main flow velocity. The density and velocity are taken from the CFD-solution at a parallel plane to the endwall which is placed at a spanwise wall distance of z/z_{ms} = 0.0067. The specific heat capacity at constant pressure $c_{p,m}$ and the Prandtl number are set for air to 1005.14 J/kgK and 0.710, respectively. The dynamic viscosity μ for air is evaluated as function of gas temperature and pressure based on a formulation proposed by Lemmon and Jacobson [78].

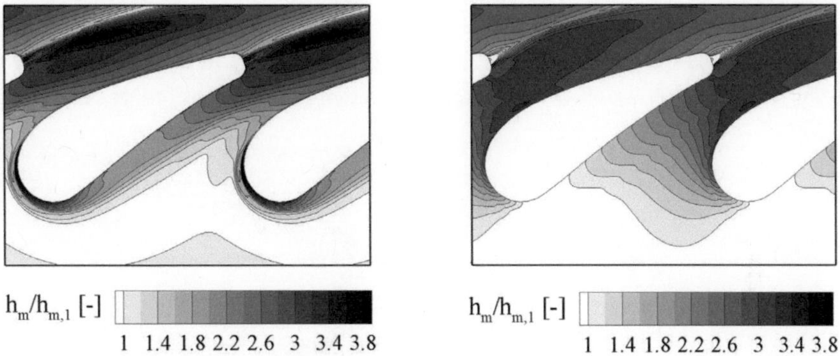

Figure 6.4: Main flow heat transfer coefficient $h_m/h_{m,1}$ from CFD-simulations (left) and correlation-based heat transfer coefficient $h_{m,corr}$ (right) for the vane's operating point at $\beta_{1,ms} = 94.8°$

Comparing the spatial distributions of the heat transfer coefficient shown in Fig. 6.4, it can be observed that heat transfer coefficients derived from the CFD are an order of magnitude lower than what is estimated by the correlation defined by equation (6.8). The CFD-computed distribution shows a distinct heat transfer gradient across the passage due to the acceleration and is significantly influenced by the near-wall flow field. For the endwall regions that are dominated by the secondary flow, the empirical relationship tends to visibly underpredict the heat transfer.

Significantly enhanced heat transfer is found around the leading edge. This is mainly attributed to the two branches of the horseshoe vortex which develop in the leading edge - endwall corner and are transported along SS and PS into the adjacent passages. Here, the heat-transfer enhancement is observed with $h_m/h_{m,1} = 3$. Less increased heat transfer ratios of 1.9 are found along the pressure side which are mostly caused by the near-wall cross flow gradient. Downstream of the trailing edge, increased increased heat transfer is observed which can be attributed to the fillet's influence enhancing the vortical motion of the passage vortex. Due to that, very high heat transfer rates of up to 3.6 can be observed in the wake region downstream of the fillet.

Concluding from the observations given above, it is not suitable to use equation (6.8) for estimating platform heat transfer. Depending on the position on the endwall, it either neglects or considerably underestimates heat transfer enhancement caused by the near-wall cross flow gradient and the secondary flow vortices. Moreover, this counts also for the cases with increased airfoil loading where the vortical motion of the secondary flow is even stronger. Due to that, additional CFD-computations are performed for the both loading cases with increased incidence ($\beta_{1,ms} = 75.0°$ and $49.4°$). The results can be found in the appendix A.3. The distributions show similarly enhanced heat transfer near the leading edge and downstream of the vane compared to the reference case. The spatial extent of these wall region is slightly enlarged. Downstream of the fillet, the wake region near the endwall shows significantly increased heat transfer with variation of loading. These effects are mainly attributed to the higher profile turning.

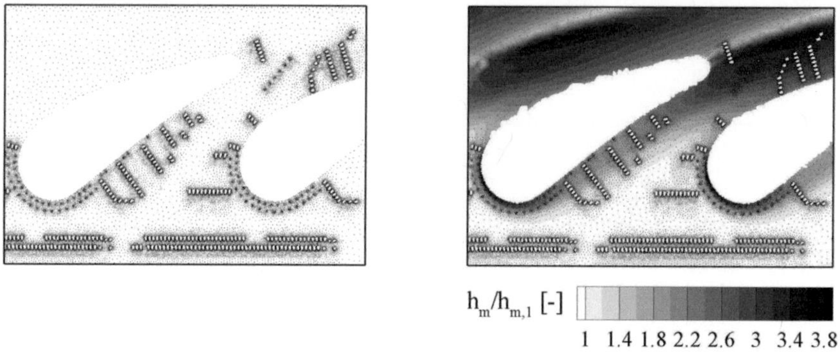

Figure 6.5: FE-nodes of endwall surface (left) and mapped heat transfer coefficients onto nodes for the operating point of $\beta_{1,ms} = 94.8°$ (right)

The computed heat transfer coefficients are spatially interpolated onto the FE-nodes of the endwall surface. The resulting pattern is displayed in Fig. 6.5 for the vane's operating point. After importation into the FEA-solver it is applied as boundary condition on the endwall surface. Adiabatic boundary conditions are imposed on the sidewalls of the endwall plate and on the coolant supply walls. On its bottom wall, the temperature was set to be the plenum temperature which was measured by a thermocouple installed some distance beneath the hole entries.

For the heat transfer coefficient inside the cylindrical part of the cooling holes h_h, the Dittus-Boelter relation for a fully-developed turbulent flow inside a pipe as defined by equation (6.6) was used. At the hole entrance, a correction factor is applied to account for the heat transfer enhancement caused by the funneling the coolant into the hole entrance. A range of 1.4 to 1.53 for the factor is used which depends on the length-to-diameter ratio L_h/D of the hole. This relationship is valid for Reynolds numbers Re_D based on the tube diameter between 10^4 and 10^5 and Prandtl numbers of 0.5 ... 10.

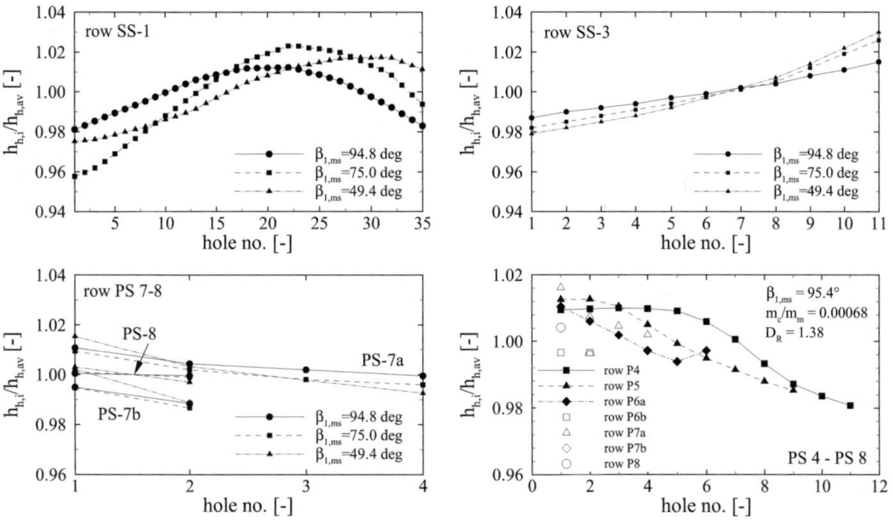

Figure 6.6: Normalized heat transfer coefficient $h_{h,i}/h_{h,av}$ for the cylindrical part of the holes of the investigated film rows SS-1, SS-3, PS-7/PS-8 and multiple row injection with PS-4 to PS-8 based on the correlation given in equation (6.6)

Moreover, the varying pressure ratio is influenced by the cascade loading which also alters the distribution of cooling flow through the holes. This attributes to the heat transfer situation inside the holes being influenced by the hole placement but also on profile loading. Following figure 6.6 depicts this effect on the local heat transfer coefficient $h_{h,i}$ for some selected rows. It is normalized with the pitchwise-averaged heat transfer coefficient for each row.

The results for the inlet row SS-1 reveal a potential flow effect on the incoming flow caused by the profile leading edge. In the wall regions where the potential flow effect takes effect, it causes a decrease in flow velocity along with an increase in static pressure. With that, the hole mass flow and hence the heat transfer coefficient decreases about 3 %. The maximum with about 1 % above the averaged row heat transfer is found around hole no. 19. With further increase of loading, the stagnation point shifts downstream of the pressure side which results into lower static pressures at the hole exits near the leading edge due to enhanced flow acceleration. Therefore, the maximum moves down the row toward greater hole numbers and is found at the hole 31 with slightly increased values of 2 % above the averaged value.

Slightly increasing heat transfer coefficients are observed across the row SS-3. An asymmetric evolution of the heat transfer coefficient can be seen around the hole 7. The variation across the row is obtained to be less than ± 2 % which gives only small increase for higher cascade incidence. Only small differences are seen for PS-7 / PS-8 with the heat transfer coefficient slightly dropping about 1 % across the row. Only negligible changes are observed when the incidence is increased.

For the multiple row injection with group PS-4 to PS-8, the estimated heat transfer distribution inside the holes for $\beta_{1,ms} = 94.8°$ is displayed in Fig. 6.6. The heat transfer distribution for the two loading cases can be found in the appendix A.3. Maximal variation of the averaged heat transfer coefficient is found to be about 1 %. The differences across the individual rows do slightly increase but not exceed 6 % with increased airfoil loading. The rows PS-4 and PS-5 seem to be more influenced compared to rows PS-7 or PS-8. This may be attributed to their endwall location closer to the profile stagnation point. The rows which are placed in less accelerated flow regimes as near the vane throat are less influenced.

The heat transfer variation across the film rows is small and far below typical measurement uncertainties of about 10 %. Thus, only row-averaged heat transfer coefficients are applied as boundary conditions for the cooling holes. The heat transfer coefficient inside the diffuser section is estimated similarly to the flat plate FE-model. Herein the decrease of coolant flow velocity is considered by the approach outlined in section 6.2.1. For the endwall case, all holes are equipped with a single-stage diffuser. Therefore, a constant diffuser area ratio of 4.33 is used for each film hole. With that, the reduction factor that is applied to the heat transfer coefficient is 0.31.

6.3 Heat Loss inside Cooling Holes

In order to determine adiabatic film-cooling effectiveness, adiabatic wall temperatures and the coolant temperature at the hole exit must be known. As a result of the inverse coolant temperature ratio, the heated coolant adds several heat load diffusing across plenum and cooling hole walls. This causes the discharge temperature of the injected film jets to be significantly reduced.

The heat loss and consequently the resulting coolant-temperature drop across the holes, is determined using a simple one-dimensional heat balance. With the calculation of the total heat flux magnitude through the plenum walls $\dot{Q}_{pl}$, the temperature change of the measured plenum temperature at the hole entry can be determined. If also the total heat flux across the cooling hole walls $\dot{Q}_h$ for each hole section is known, its temperature drop can also be calculated. In combination with the coolant mass flow rate measured for a specific row test case and the specific heat capacity $c_{p,c}$, the discharge temperature of the cooling gas is determined following

$$T_{c,dis} = T_{pl} + \frac{(\dot{Q}_{pl} + \dot{Q}_{h,cyl} + \dot{Q}_{h,fan})}{\dot{m}_c c_{p,c}} \tag{6.10}$$

with T_{pl} the plenum temperature. The magnitude of coolant-temperature drop depends on the supplied coolant mass flow rate and its share between the cooling holes. This influences the convective heat transfer inside the film holes. Additionally, slight changes in the temperature ratio between the coolant and main flow may also attribute to the heat loss. The calculations are performed for each reported film-cooling test case. The estimated temperature drop of the cooling flow due to heat loss varies between -0.3 ... $-0.8\,K$. It is different for each film-cooling row and depends primarily on the coolant-to-main flow temperature difference and the supplied mass flow rate.

6.4 Data Post Processing

The post-processing of the FE-computations to derive the adiabatic wall temperatures includes several steps. A schematic of the used methodology is outlined in Fig. 6.7. Prior to the computations, the FE-model is prepared for each film row test case by providing the gas path heat transfer coefficients and the heat transfer coefficients inside the holes. The boundary

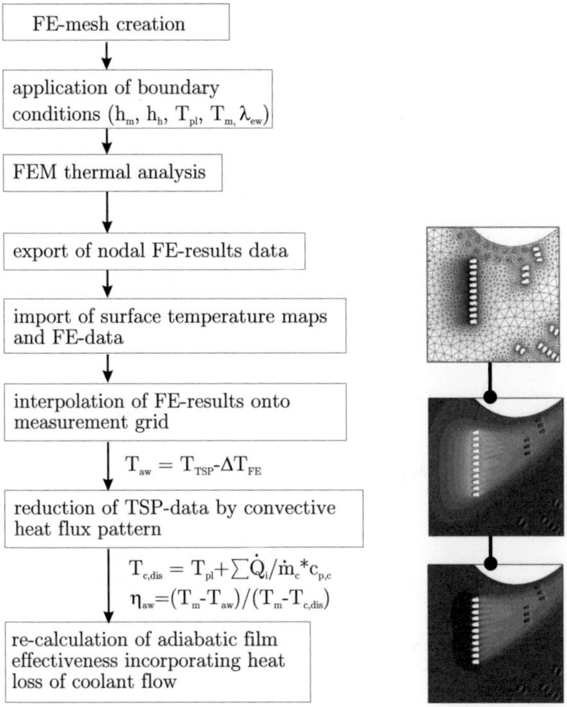

Figure 6.7: Data reduction method for TSP-measurements using conductive heat flux data from FEM-computations

conditions are applied on the mesh nodes of each model section and is imported during the solver pre-processing. The thermal analysis is initiated by applying the measured main flow temperature onto all nodes of the model. After defining the material constants for the used test plate material like thermal conductivity, density, Young's modulus and Poisson ratio, steady state heat transfer calculations are performed using the commercial Finite-Element solver ABAQUS™.

The results for the nodal temperatures as well as the heat fluxes on the endwall and for each hole section are exported. The data is combined with the nodal gas-path heat-transfer coefficients along the connectivity matrix for the triangular mesh elements of the endwall surface. After checking for consistent coordinate systems, the conductive heat-flux pattern is mapped onto the measured surface temperature distribution using a linear interpolation scheme. The measurement grid contains 1600 *pixel* x 1200 *pixel* at the CCD-sensor's full resolution which equals to a spatial resolution of 0.13 *mm* between two pixel. The FE-mesh resolution is slightly coarser yielding element sizes between 0.4 *mm* - 0.6 *mm* for the hole exit regions. The refinement of the FE-mesh resolution to achieve comparable element sizes would result into multiplying the total amount of the model's tetrahedral cells. Thus, it requires immense computational resources that were not available. Nevertheless, an element size ratio of 0.32 between both data sets is found to be sufficient enough to achieve acceptable results with the interpolation scheme. After successful mapping of the conductive heat-flux pattern, the computed wall-temperature difference is used to eliminate the heat conduction effects in the experimental data.

The observed heat conduction effects mainly occur several diameters upstream of the holes and distribute in lateral direction where the film-cooling jets do not affect the surface temperatures. This means that the increase in wall temperature in those regions is a direct result of the heat conduction process through the non-perfect adiabatic test plate. It may not be attributed to changed convective boundary conditions due to the presence of the film layer. Applying a one-dimensional heat flux balance on surface elements in such endwall regions where no coolant is present, one can state that

$$\dot{q}_{conv} = \dot{q}_{FE,cond} = h_m \left(T_w - T_{aw}\right). \tag{6.11}$$

The latter is only valid as long as the energy balance is limited to endwall regions where the heat transfer enhancement by the presence of the coolant can be neglected and the change in surface temperature is only caused by wall heat conduction. Thus, it can be applied only to surface elements upstream of the film-cooling hole exits. The adiabatic wall temperature can be calculated based on the specific wall heat flux that is derived from the FE-analysis as

$$T_{aw} = T_w - \Delta T_{FE} \tag{6.12}$$

$$\Delta T_{FE} = \left(T_{aw} - T_w\right) = -\frac{1}{h_m}\,\dot{q}_{FE,cond} \tag{6.13}$$

with h_m the gas path heat transfer coefficient and $\dot{q}_{FE,cond}$ the specific heat flux due to wall conduction. For the application of the correct heat transfer coefficient during data reduction, it must be distinguished between the experiments obtained on the flat plate and the endwall film-cooling measurements in the turbine cascade. The issue is easily resolved for the flat-plate film-cooling experiments. Here, a constant gas path heat transfer coefficient used which is derived from equation (6.1). It turns out to be a bit more complicated for the endwall film-cooling experiments. As shown, the heat transfer coefficient on the endwall shows distinct axial and lateral differences and can not be represented by a single value. Therefore, the computational results are used according to equation (6.7). With the adiabatic wall temperature and the coolant exit temperature estimated from the heat loss across the holes, the adiabatic film-cooling effectiveness is finally calculated with regard to equation (2.7).

6.5 FE-Results for Flat Plate Film-Cooling

For the flat-plate film-cooling experiments, the conductive heat flux pattern obtained by the thermal analysis are compared to the experimental surface temperature fields and the adiabatic wall case after the FE-data reduction of the measurements. The latter was applied to all flat plate test cases. The results are shown in Fig. 6.8 for both film hole geometries. A test case with a blowing rate of $M = 0.75$ and CO_2 injection was selected.

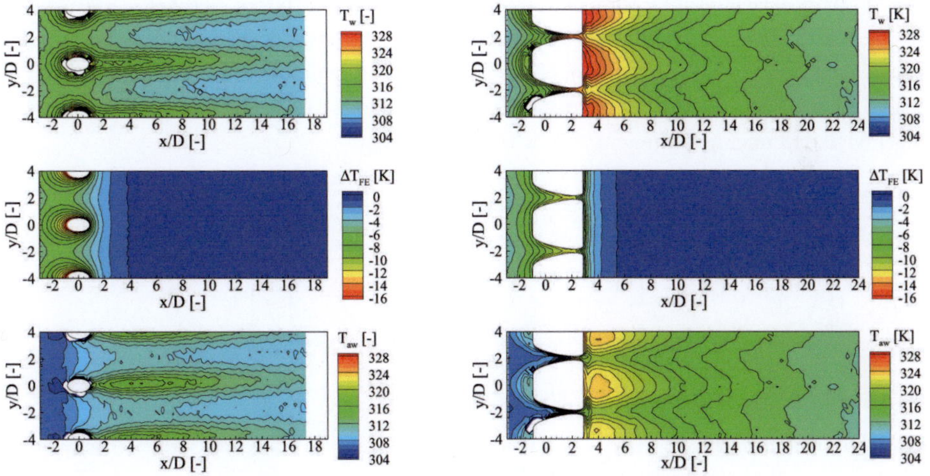

(a) Cylindrical Holes, $M = 0.75$, $D_R = 1.37$

(b) Laid-Back Fan-Shaped Holes, $M = 0.75$, $D_R = 1.37$

Figure 6.8: Surface temperature field obtained via TSP, computed FE-wall temperatures due to heat conduction and resulting adiabatic wall temperatures for the test case with $M = 0.75$ and $D_R = 1.37$ for the cylindrical and shaped holes

Observing the measured wall temperature field for the cylindrical holes, elevated surface temperatures are visible upstream of the holes and in lateral direction between the holes. The extent of the computed, conductive heat flux asymptotically decrease with increasing normalized streamwise distance x/D. The propagation of the conduction pattern in spatial direction is determined at the point where the temperature ratio T_w/T_m shows less than 0.5 % difference. It shows a remarkable extent in downstream direction and approaches the static main flow temperature at $x/D = 4$. With respect to the downstream hole edge the conductive wall temperature declines over two hole diameter. In upstream direction the pattern spreads further outwards and the temperature gradient declines over more than 6 hole diameters. The surface temperature is observed to be 2.6 % higher on the leeward hole edge than closely downstream of the hole exit. Between two cooling holes, the conduction pattern reveals decreasing surface temperatures with its minimum located at $y/D = 2$ halfway between the holes.

Although the primary heat conduction effects are eliminated by the applied FE-correction, still some surface regions with elevated temperatures are observed. This phenomenon mainly occurs in the wall region between the film-cooling jets for streamwise distances of $3 < x/D < 7$. The lateral wall heat conduction is caused by the inverse coolant-to-main flow temperature ratio with a coolant discharge temperature that is about 20 - 30 K higher than the static main flow temperature. With a lateral temperature gradient between the hot jet core and the cold wall areas where no coolant is present, conductive heat fluxes were initiated inside the test plate (see also Fig. 6.9). The largest increase in wall temperature between the jets occur at the streamwise position where the centerline temperature evolution shows the highest film temperatures. The effect diminishes further downstream of the holes due to a reduced temperature gradient. This results from decreasing film temperatures inside the cooling-jet core in streamwise direction which is due to mixing and interaction with the main flow.

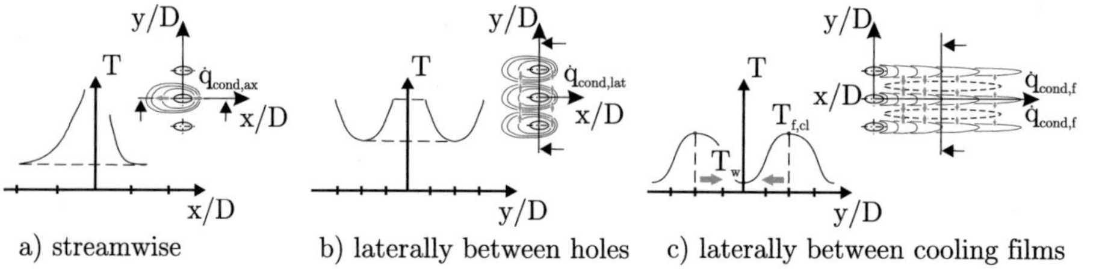

Figure 6.9: Schematic of heat conduction effects observed in TSP-surface temperature measurements

Considering the case with laid-back fan-shaped holes, the extent of the conduction pattern is similar in upstream and downstream direction compared to the cylindrical holes. It sharply declines in streamwise direction to reach the main flow temperature level at $x/D = 6$. With respect to the downstream diffuser edge, the conductive wall temperature declines over two hole diameters. The extent in upstream direction is found to be larger. It spreads over more than 5 hole diameter. Here, the driving temperature difference for the upstream conduction process is larger than in streamwise direction. The temperature difference is defined by the wall temperature close to hole exit and surface temperature far off the injection location which equals the static main flow temperature.

Due to the large diffuser angle and the entire diffuser portion of 4D, the width of the hole breakout is quite large measuring about 1.8 hole diameters. It is therefore 3.6 times greater than the elliptical shape of the cylindrical holes. With the reduced material thickness between the hole exits and considering that the holes are placed at a relatively small pitch ratio of $p/D = 4$, the temperature drop in lateral direction is far less intense. This indicates enhanced heat conduction with much higher conductive wall temperature differences.

On the contrary, the small pitch ratio along with the diffusion inside the hole causes the separate film jets to be broadened which results in enhanced lateral spreading. It further leads to partial merging of the films at an early stage in streamwise direction. This forms a closed protective layer on the cooled surface which would in turn significantly reduce the lateral heat conduction between the film jets. As visible in the corrected temperature distribution, the effect of lateral heat conduction is less pronounced for the used shaped holes. Hence it plays a minor role in affecting the surface temperature distribution and will be disregarded for further considerations.

Using the presented FE-approach on the results for the flat-plate film-cooling, the major heat conduction effects visible in the temperature distributions could be evaluated. Still some side effects concerning lateral heat conduction are observable between the film-cooling jets that could not be captured. This issue is mostly relevant for cylindrical holes due to the separate jet structure which alters the lateral temperature distribution and consequently laterally-averaged film effectiveness. For fan-shaped holes, those effects are less important as long as the combination of pitch ratio and diffuser opening angle causes merging of the separate cooling films to form a closed coolant blanket. The observed lateral heat conduction is mainly dependent on the streamwise temperature evolution inside the cooling jets. Therefore, the corresponding film temperatures are dictated by the jet flow field and the mixing process with the main flow. To integrate these heat conduction effects into the FE-data reduction process for further film-cooling investigations, an advanced approach combining solid-state simulations with CFD-methods (denoted as conjugate heat transfer) is suggested.

6.6 FE-Results For Endwall Film-Cooling

The measured spatial wall temperature distribution, as depicted for the row SS-3 in Fig. 6.10 a), shows elevated surface temperatures upstream and sideways of the film row forming a characteristic heat conduction pattern around the holes. The extent in axial direction upstream of the holes is quite large until $x/D = -24$ whereas the conductive temperature gradient degrades more swiftly in lateral direction within a distance of about 5-8 hole diameter. Downstream of the holes, the effect is less pronounced leading to a slightly quicker decay of the surface temperature within 6 hole diameters. The propagation of the heat conduction pattern in spatial direction is determined at the corresponding locations where the temperature ratio T_w/T_1 shows less than 0.5 % difference.

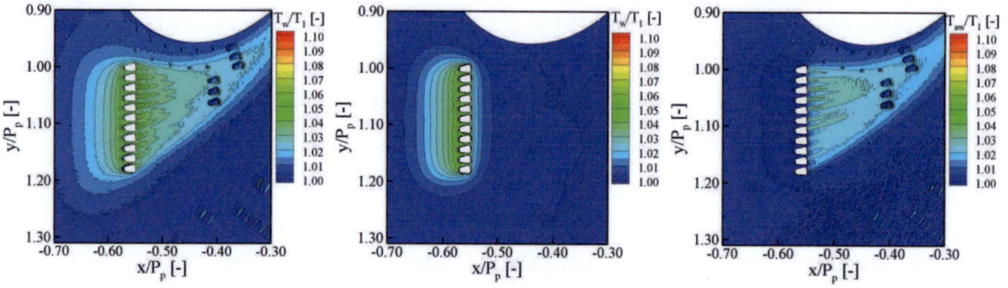

(a) row SS-3, M = 1.60, D_R = 1.40

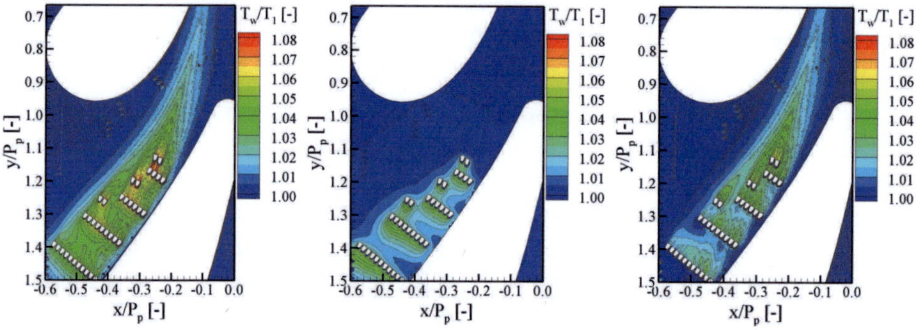

(b) rows PS4-PS8, $\dot{m}_c/\dot{m}_m$= 0.007, D_R = 1.37

Figure 6.10: Measured wall temperature distribution (left), FEM-wall computed conductive wall temperatures (center) and revised surface temperature field (right) for endwall film-cooling through row SS-3 (a) and multiple row injection through PS-4 to PS-8 (b), vane's operating point at $\beta_{1,ms} = 94.8°$ (data normalized with cascade inlet temperature T_1)

The observed heat conduction is mainly driven by the temperature difference between the coolant and mainstream with the heat load across the plenum and the cooling hole walls additionally attributing to this effect. The heated coolant flowing through the holes initiates a conductive heat flux pattern on the endwall surface propagating in upstream and downstream direction. The temperature gradient between the hot film layer and the colder main flow further drives lateral, conductive heat fluxes across the outer edges of the film layer. Heat conduction between individual jets inside the film layer are of minor importance since the film streaks merge very early, not far downstream of the hole exit. This is due to the large diffuser angle and moderate hole pitch-ratio.

The propagation of the heat conduction in upstream and in lateral direction around the holes is matched qualitatively by the FE-thermal analysis. The decay of the conductive wall temperature in upstream and lateral direction is reproduced quite well. Near the hole exits, the change in surface temperature is slightly overestimated which results into very small film effectiveness value in those areas. The observed endwall conduction effects are very similar to what had been observed in the flat plate film-cooling experiments. To assess these effects for further endwall cooling investigations, conjugate heat-transfer simulations must be applied to solve the flow field along with the thermal distribution inside the solid.

Similar trends in the evolution of the conductive wall temperature pattern are observed for the other investigated single rows PS-7/PS-8, SS-1 and for the test case with multiple row injection of the rows PS-4-to PS-8. Observing the results from the FE-computations for the multiple-row test case, as shown in Fig. 6.10 b), the developed conduction pattern is similar from row to row while its entire upstream and lateral extent is more pronounced than for single-row injection. Nevertheless, the conduction patterns that are developed around PS-5 up to PS-8 unite partially between the rows due to short row spacing and staggered hole positions. As a result of that, the FE-computed wall temperature steadily decreases in streamwise direction. Until it rises up again at the point where the pattern of the next row begins to influence the wall temperature distribution. Similar for all rows, the heat conduction pattern degrades swiftly within 12 hole diameter in lateral direction. Downstream of the hole injection, the temperature gradient is similar to what was observed for the single-row injection. It shows a strong decay of the conductive heat flux in streamwise direction within about two hole diameters. Enhanced propagation of the pattern is found in upstream direction which slightly varies from row to row being influenced by the conduction pattern of the neighboring rows.

Observing the results from the FE-computations, it can be seen that the major effects as the observed heat-conduction pattern around the holes are reproduced and show similar extent in lateral and upstream direction with regard to the hole exits. The magnitude of the temperature increase caused by heat conduction is slightly overestimated in the immediate vicinity of the hole exits which results into reduced adiabatic film effectiveness.

7 Flat-Plate Film-Cooling

7.1 Wall Boundary Layer

Prior to the film-cooling experiments, the main flow boundary layer in the test duct was measured. A pitot-probe was used to obtain the total pressure-loss distribution across the channel height and 1D-hot wire probe to examine the turbulence intensity. The boundary layer profiles are determined at a constant main flow velocity of $u_m = 22.9\,m/s$. The probes were inserted through the upper wall of the test duct at two different locations upstream of the film-cooling holes at $0.355\,m$ (ML 1) and $0.180\,m$ (ML 2), respectively. For the cylindrical

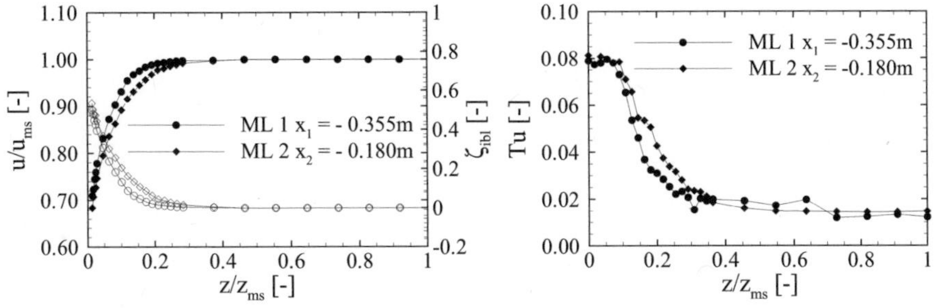

Figure 7.1: Normalized velocity profiles u/u_m, total pressure loss coefficient ζ_{ibl} and turbulence intensity Tu for two different measurement stations at x/D = -118 and -60 upstream of the film holes at constant main flow velocity of $u_m = 22.9\,m/s$

holes, these normalized distances equal to x/D = -118 and -60. Due to the diffuser hole's larger diameter for the metering section with $D = 5\,mm$, the relative positions turn out to be smaller with x/D = -71 and -36. The insert locations for the probes are selected in order to place the sensors as close as possible to the film-cooling holes without creating any disturbances for the optical access. The total pressure distribution of the test duct boundary layer was determined at 27 sections along half of the channel height. The probe was manually traversed observing the centerline section of the test duct. As a result of the finite extension of the probe head with a diameter of $1.1\,mm$, the minimum wall distance is $0.75\,mm$.

The normalized velocity profiles u/u_m are calculated from the total pressure measurements. Due to limited spatial resolution, the velocity profiles are extrapolated toward the wall using

location		u_m	δ_{99}	δ^*	Θ	H	Ma	Re_x
	x [m]	[m/s]	[mm]	[mm]	[mm]	[-]	[-]	[-]
ML 1	-0.355	22.9	11	1.228	0.959	1.281	0.066	$2.78{\cdot}10^5$
ML 2	-0.180	22.9	15	1.665	1.295	1.286	0.066	$5.49{\cdot}10^5$

Table 7.1: Summary of the wall boundary layer parameter measured for the two measurement stations upstream of the film hole row

the 1/n-power law of a turbulent boundary layer. Detailed description of the data analysis and equations for the boundary layer parameter including the definition for the shape factor can be found in section 5.1. The results of the boundary layer measurements are shown in Fig. 7.1 for the two insert locations upstream of the film holes. Observing the obtained velocity profiles, differences are visible between the two measurement locations which indicate considerable growth of boundary layer with reduced distance from the hole exits. This is also confirmed by the velocity boundary-layer thickness, the displacement and momentum thickness which deliver increasing magnitudes in streamwise direction. The streamwise increase in boundary-layer thickness was found to be 33.3% for $u_m = 22.9\,m/s$. The displacement thickness increases similarly from $\delta^* = 1.23\,mm$ at ML 1 upstream of the holes to $\delta^* = 1.67\,mm$ at ML 2. Since no flow manipulation was applied in terms of trip wires or boundary layer extraction, the flow passes a undisturbed transition process through the wind tunnel components. The calculated shape factors, as given in Table 7.1, identify hereby a fully-developed turbulent boundary layer for all test cases.

According to Schlichting [114], the displacement thickness of a turbulent boundary layer with no pressure gradient follows the $1/7^th$ polynominal law described by $\delta^* = \delta/8$. In this case, this parameter is calculated at ML 1 for a main flow velocity of $u_m = 22.9\,m/s$. It measures 1.875 which is in good agreement with the data obtained from the boundary layer profiles, also seen in Table 7.1.

Observing the total pressures losses, a steadily growing loss coefficient with decreasing wall distance can be found. The highest pressure loss is found close to the wall with $\zeta_{ibl} = 0.50$ for ML 1 and $\zeta_{ibl} = 0.53$ for ML 2. This indicates a slight loss increase in downstream direction. The profile is found to be slightly thickened and extended in wall-normal direction due to growing friction losses. The span-wise extent of the region with high losses corresponds well with the boundary layer thickness obtained from the velocity profiles. It extends about 28 % of mid-channel height at ML1 and $z/z_{ms} = 0.37$ for ML 2.

The turbulence intensity of the main flow is obtained using the Dantec CTA-System (for details see section 4.3). The turbulence intensity at channel midspan is fairly constant over a large portion of the test section height. At a wall distance of $z_w/z_{ms} = 0.46$ it starts to rapidly increase to values of 8% close to the wall due to enhanced turbulence inside the boundary layer. It is found to be nearly constant for the last 10 % of the wall distance.

The turbulence intensity at channel center is obtained to be uniformly distributed for the measurement location ML 2 with values of 1.44 % in the core of the main flow. At the location ML 1, slight non-uniformities in the turbulence intensity is found. The curve shows an equivalent trend along the channel height compared to ML 2 but gives slightly more pronounced variation of the turbulence intensity. This is may be attributed to an influence of the contraction nozzle which is located 0.48 m upstream of the measurement location ML 1. While the nozzle is used to transfer the cylindrical cross-section into the duct's rectangular shape, the sharp corner edges inside the nozzle alter the wall pressures. Hence, small corner vortices are initiated which are transported into the test channel. Due to interaction of these vortices with the primary flow results into streamwise mixing, their influence on the turbulence distribution is just recognized at ML 1 and diminishes further downstream.

7.2 Film Effectiveness for Cylindrical Holes

Film-cooling experiments have been carried out for injection of CO_2 and air, thus applying varying coolant-to-main flow density ratios. Surface temperature fields are obtained for a blowing rate range of $0.35 < M < 1.5$. Hereby, the coolant gas is heated up to provide a coolant-to-main flow temperature ratio of $T_R = 1.12$. Though it lies within the working temperature range of the TSP, it establishes a large temperature spread which provides sufficient resolution for the TSP-system. This temperature ratio is maintained for all test cases. It results in a density ratio for air of $D_R = 0.93$ and for carbon dioxide of $D_R = 1.37$ with the main flow temperature held constant at 30°C. During the data post-processing, the surface temperature fields are corrected for the observed heat conduction effects. The adiabatic film effectiveness is then determined according to equation (2.7). The applied FE-method is presented in detail in section 6.

TSP had been applied to the film-cooled test plate in a region of 180 mm x 110 mm downstream and 50 mm upstream of the holes. Image series are taken for a single camera position without traversing. Due to the restricted field-of-view of the camera-system, the streamwise measurement region is limited to the near-field of the injection. It covers five out of the seven film holes. As a result, the observed test plate area for the cylindrical holes extends in axial to -2 $< x/D <$ 18 and in pitchwise direction -9 $< y/D <$ +9. Therefore images at a single streamwise positions are acquired. Due to this, the obtained measurement data covers the film effectiveness decay in the near-field of the film injection excluding its asymptotic approach farther downstream. Traversing of the camera to acquire images at different streamwise distances is not considered. This approach reduces the experimental efforts and avoids measurement errors introduced by overlapping image regions.

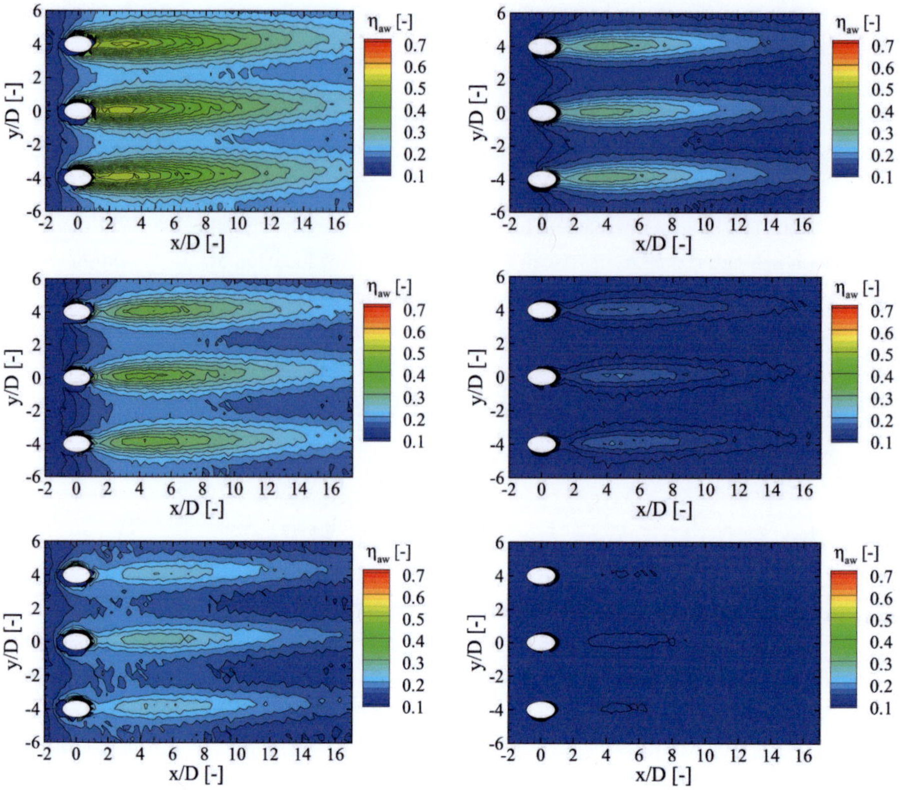

(a) $D_R = 1.37$, M = 0.50, 0.75 and 1.00

(b) $D_R = 0.90$, M = 0.53, 0.71 and 1.00

Figure 7.2: Local film effectiveness η_{aw} obtained for $M = 0.5$, 0.75 and 1.0 with CO_2 and air injection, $Re_{m,D} = 4200$, $u_m = 22.9\,m/s$, $Tu = 1.5\,\%$

The local film effectiveness distributions, shown in Fig. 7.2, had been analyzed for the three center holes of the row. Centerline effectiveness was extracted from the TSP-data sets on a pixel-by-pixel basis and is interpolated for each streamwise distance over a range of 10 pixel. Lateral averaging is carried out across four pitches (5 holes). The measurement signals inside the hole exits are neglected for the data reduction since no TSP data is available in this region. Also the embedded surface thermocouples reduce the TSP-luminescence signal so that the surface temperature and hence the film effectiveness is significantly altered in these areas resulting into un-physical high or low values. Therefore the wall regions around the thermocouple locations $\pm 0.9D$ are excluded from the data reduction.

Centerline and laterally averaged film-cooling results are shown in Fig. 7.3 versus the

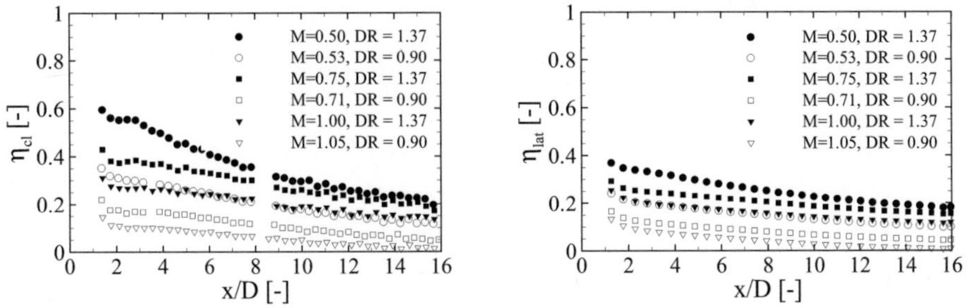

Figure 7.3: Centerline η_{cl} and laterally averaged film-cooling effectiveness η_{lat} obtained for $M = 0.5$, 0.75 and 1.0 with CO_2 and air injection, $Re_{m,D} = 4200$, $u_m = 22.9\,m/s$, $Tu = 1.5\,\%$

dimensionless streamwise coordinate x/D at three selected blowing rates 0.5, 0.75 and 1.0. Highest film effectiveness values are observed at the lowest blowing rate $M = 0.50$ with CO_2 and density ratio $D_R = 1.37$. With increase of the blowing rate to $M = 1.00$, centerline film effectiveness is significantly reduced due to the increased jet momentum. Centerline film-cooling effectiveness shows values of 0.55 close to the injection location for $M = 0.50$ which is reduced to 0.28 for $M = 1.00$. With increased distance downstream, steadily decrease in film-cooling effectiveness is observed due to the mixing of the jet with the cross flow. The results for air injection with a density ratio of 0.9 indicate 30-35 % lower film effectiveness even for the case with low blowing rate. These effects correspond with the film jet behavior observed from the local film effectiveness distribution.

Lowest values of centerline effectiveness are observed for a blowing rate of $M = 1.00$ and air injection with low density ratio ($D_R = 0.93$) which indicates detachment of the cooling jets from the surface. The same trend can be found for the laterally-averaged film effectiveness. At a low blowing rate of $M = 0.50$ and high density ratio, highest film effectiveness values are found which reduce with increasing downstream distance. Reduced film effectiveness is obtained for air injection due to the increased jet momentum which causes the jets to lift off from the surface and penetrate into the cross flow. For a particular blowing rate, the injection of CO_2 produces higher film effectiveness compared to air due to the more dense cooling jets. It causes lower momentum which forces the jets to stay closer attached to the surface. Using air as coolant shows reduced film effectiveness due to the higher momentum flux ratio. It results into increased tendency of film detachment from the surface reducing their protection capabilities. These findings coincide with the behavior described by Ekkad et al. [31] and Sinha et al. [121].

7.3 Comparison to Literature Results

Film-cooling injection for a single row of inclined holes had been studied by many investigators in the past. So, significant differences can be observed in the experimental setup and the used geometrical parameters Several references had been selected to compare their results with the present measurements in terms of centerline film effectiveness. Table 7.2 summarizes the fluid and geometrical boundary conditions of the present work and the previous investigations as far as available from the publications.

(a)	P_h/D [-]	L_h/D [-]	α [°]	D [mm]	N [-]	cooling gas
present study	4	4.6	35	3	7	CO_2, air
Ekkad [31]	4	4.6	35	4.75	7	CO_2, air
Lutum [90]	2.5	5	35	4	7	air
Pedersen [102]	2.5	40	35	11.7	5	CO_2, air, ...
Sinha [121]	3	1.75	35	12.7	7	air

(b)	M [-]	D_R [-]	I [-]	Tu [%]	u_m [m/s]	δ^*/D [-]
present study	0.5, 1.0	1.37	0.19, 0.73	1.5	22.5	0.5
Ekkad [31]	1	0.98	1.02	8.5	8.35	0.267
Lutum [90]	0.52, 1.16	1.15	0.24, 1.16	3.5	115	0.5
Pedersen [102]	0.515, 1.06	1.18	0.224, 0.936	0.4	27.7	0.16
Sinha [121]	0.5, 1.0	1.2	0.21, 0.83	0.2	20	-

Table 7.2: Comparison of geometrical (a) and experimental boundary conditions (b)

The study from Ekkad et al. [31] was selected because the author had used air and CO_2 as foreign gas injection but unfortunately at a higher free-stream turbulence intensity of $Tu = 8.5\,\%$. Centerline film effectiveness data is available for simple angle holes and holes with a compound angle of 45° and 90°, respectively. Effectiveness data for the simple angle holes is limited to conditions with a blowing rate $M = 1$ and a density ratio $D_R = 0.98$. Lutum et al. [90] investigated the influence of the hole length-to-diameter ratio on cylindrical holes. The parameter L_h/D was varied from 1.75 to 18. Their test case of $L_h/D = 5$ matched the conditions of the presented experiments best and was therefore chosen for the comparison.

Pedersen et al. [102] and Sinha et al. [121] studied film-cooling injection on a single row of inclined holes at very low free-stream turbulence (Tu = 0.4 and 0.2 %, respectively). While Pedersen et al. [102] investigated film-cooling injection through very long holes ($L_h/D \sim 40$) using the heat-mass transfer analogy in applying large differences in density ratio from $D_R =$

0.75 to 4.17, Sinha et al. [121] used very short holes with $L_h/D = 1.75$ instead. Two coolant blowing rates at $M = 0.5$ and $M = 1.0$ with CO_2 injection were selected for comparison. They equal to a momentum flux ratio of $I = 0.19$ and 0.73 respectively. Fig. 7.4(a) compares centerline film effectiveness for the case with low momentum ratio $I = 0.19$ and Fig. 7.4(b) shows effectiveness data for the increased jet momentum. Comparison for the low momentum case shows that the centerline effectiveness agrees well with the references in the region close to the injection location. The present data decreases at the same decay rate with increasing distance downstream of the hole-injection location. Small differences in film effectiveness at $M = 0.5$ are found which may further reduce if data would be extrapolated to larger values of x/D. Slightly lower film effectiveness is observed compared to the data of Pedersen et al. [102]

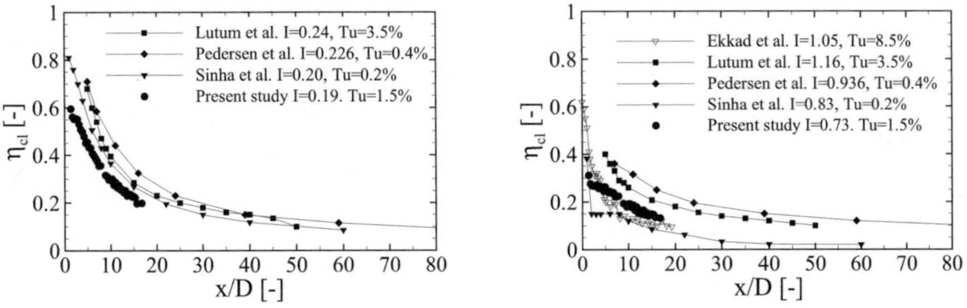

(a) low momentum flux ratio (b) elevated momentum flux ratio

Figure 7.4: Comparison of centerline film-cooling effectiveness to previous literature results for $M = 0.5$, $I = 0.19$ (a) and $M = 1.0$, $I = 0.73$ (b) with CO_2-injection, $D_R = 1.37$, $Re_{m,D} = 4200$, $u_m = 22.5\,m/s$, $Tu = 1.5\,\%$

which may be attributed to the long holes used by the authors compared to the shorter holes (L_h/D 4.6) used for the present study.

As the blowing rate and consequently the momentum flux ratio is increased, the differences in film effectiveness become more pronounced. Higher centerline effectiveness is achieved for $I = 0.73$ compared to the data of Ekkad et al. [31] and Sinha et al. [121] but agrees reasonably well for both data sets at small $x/D < 4$ close to coolant injection. Further downstream of the hole exit, lower effectiveness measured by Ekkad et al. [31] may be caused by the increased momentum flux ratio and elevated free-stream turbulence level which creates stronger mixing of the film jets with the main flow. Results from Sinha et al. [121] indicate that for short holes, the jet detachment occurs somewhat earlier which coincides with reduced film effectiveness. The investigators state that the effective injection angle is increased which causes earlier lift-off at smaller jet momentum. Compared to the present results, lower effectiveness can be found which may be attributed to the described effect. The results of the present work show lower film effectiveness values compared to the data of Pedersen et al. [102] which may be attributed to the low free-stream turbulence intensity and the extremely large hole length-to-diameter ratio. Lutum et al. [90] measured higher centerline effectiveness for the blowing rate $M = 1.0$.

The film-effectiveness data obtained on the flat plate is further compared to detailed CFD-computations done by auf dem Kampe and Völker [3] that are based on the present experimental setup. The authors provided numerical investigations of a single cylindrical holes extending one pitch on a flat plate. The flow domain comprised the main flow path, cooling hole and the supplying plenum chamber. The hexahedral, block-structured low-Reynolds grid features a high node density of $5{\cdot}10^6$ nodes to resolve the wall boundary layer of the main flow and inside the cooling hole. The computations are conducted using ANSYS CFX^{TM} 11 applying the SST- turbulence model by Menter [95]. To account for the heat-conduction effects through the perspex test plate due to its finite thermal conductivity, the authors selected a conjugate heat-transfer approach. The measured total pressure profile, the turbulence intensity profile and the total temperature of the main flow as well as the total temperature and total pressure of the coolant in the plenum are imposed as boundary conditions. Further details on the used model can be found in [3]. The comparison of the numerical results versus the experimental data for two different density ratios using either air or CO_2 at constant blowing rate is shown in Fig. 7.5.

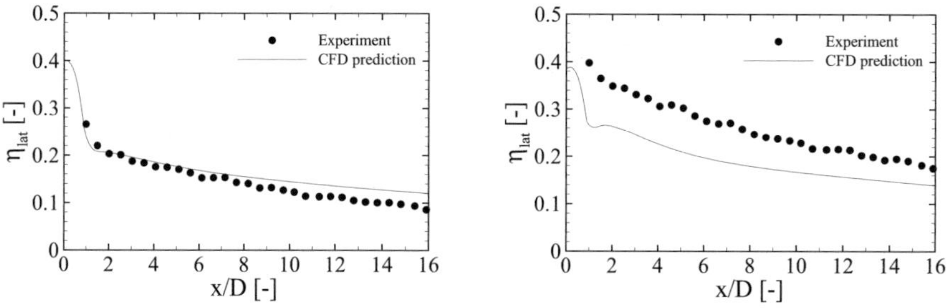

(a) $M = 0.5$, $D_R = 0.90$, coolant: heated air (b) $M = 0.5$, $D_R = 1.37$, coolant: heated CO_2

Figure 7.5: Comparison of laterally averaged effectiveness CFD vs. Experiment from auf dem Kampe and Völker [3]

The CFD-predictions of the laterally-averaged film-cooling effectiveness agree remarkably well with the experimental data of the cylindrical holes for the selected blowing rate. The curves match better for the low density ratio where the laterally-averaged film effectiveness could be predicted within a bound of ± 0.05. Deviations become slightly more pronounced with up to ± 0.1 locally for the high density ratio case with heated CO_2-injection. Despite the spatial differences, the prediction of the qualitative trend in terms of the streamwise decay of the film effectiveness agrees quite well with the experiments.

7.4 Film Effectiveness for Shaped Holes

The film-cooling investigations for the shaped holes are carried out in a similar way as done for the cylindrical holes. The difference in coolant-to-main flow density ratio is achieved by using heated carbon dioxide or air which equals to a density ratio of $D_R = 1.37$ (CO_2) or $D_R = 0.90$ (air). The temperature ratio was maintained at $T_R = 1.11$ by heating the coolant to a constant coolant temperature level despite the variation in the coolant mass flow rate. The blowing rate was varied between 0.35 to 2.0. TSP had been applied to the test plate in a region of 180mm x 110mm downstream and 40mm upstream of the holes. The same camera and illumination setup, as during the investigations with the cylindrical holes, is used. Due to the larger hole diameter and pitch of the diffuser holes, the distance between the lens and the test plate is increased to achieve a similar viewing field and to observe all hole exits of the film row. Herein the measurement region covers $-2 < x/D < 26$ in axial direction and $-10 < y/D < +10$ in pitchwise direction.

For the injection through the laid-back fan-shaped holes, the local film effectiveness distributions for three blowing rates with differences in density ratio are presented in Fig. 7.6. The blowing rate for the fan-shaped holes is calculated with the inlet cross-sectional area of the hole metering section. At low blowing rate $M = 0.50$ and air injection, lower film effectiveness of about 0.40 is obtained immediately downstream of the hole exits. Compared to the case with CO_2, the film effectiveness is higher being observed at 0.65 at the same wall region. All test cases show enhanced lateral spreading of the coolant compared to the cylindrical test case. In general, this is a result of the diffuser effect which is enhanced by large lateral opening angles of 14°. Comparing CO_2-results with the air injection case for the small blowing ratio of 0.50, slightly reduced lateral spreading can be observed further downstream of the hole injection location.

Reduction in film effectiveness occurs for the lower density ratio when the blowing rate is further increased beyond 0.50. The differences are about 8-10 % close to the hole injection and diminish steadily as the jets proceed downstream. The results indicate that the effect of the density ratio is small using fan-shaped holes. It is seems to be spatially limited to the wall region closely downstream of the hole exits.

Compared to the cylindrical holes, the effect of the density ratio is significantly smaller due to the decrease of the coolant velocity inside the diffuser parts. This affects in turn the momentum flux ratio of the coolant. So, it is reasonable that the effect of density ratio is negligible at low blowing rates but tends to increase with higher blowing rates. Lower film effectiveness at high blowing rates ($M = 1.00$) with low density ratio is observed close to the hole injection location.

The film effectiveness increases with larger blowing rate for both density ratios whereas only small changes in the lateral distribution of the coolant is observed for the case of CO_2 injection. The film effectiveness close to the hole exit shows a slightly non-symmetric distribution. This

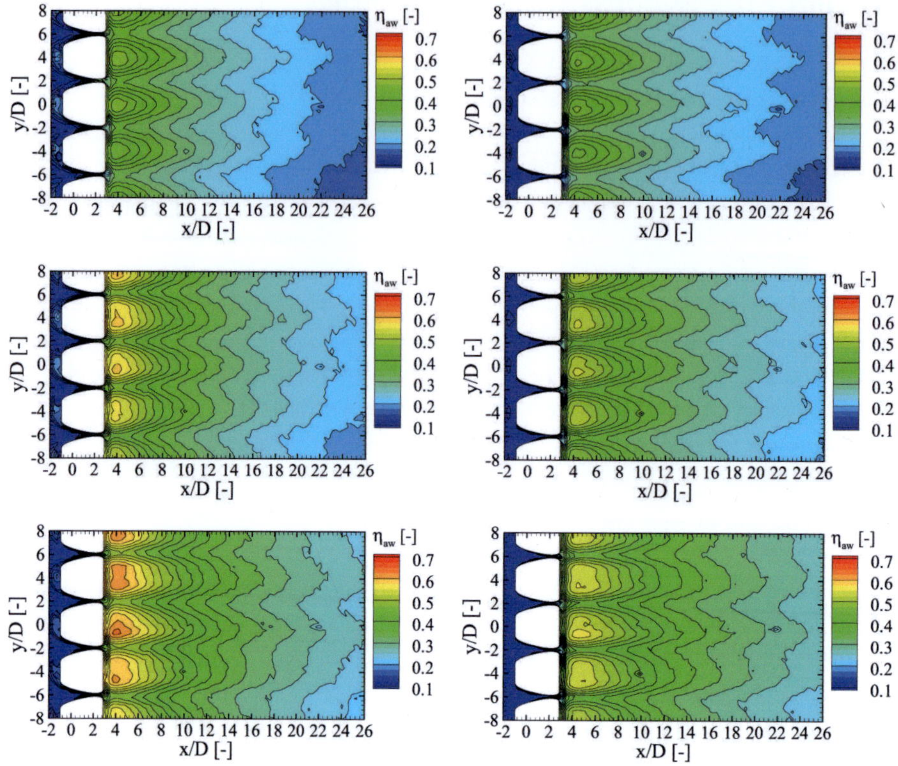

(a) $D_R = 1.37$, $M = 0.5$, 0.75 and 1.00

(b) $D_R = 0.90$, $M = 0.53$, 0.70 and 1.00

Figure 7.6: Local film effectiveness η_{aw} for a row of laid-back fan-shaped holes obtained for several blowing rates using CO_2 and air injection, $Re_{m,D} = 7700$, $u_m = 22.9\,m/s$, $Tu = 1.5\,\%$

effect becomes more pronounced as the blowing rate increases which coincides with the findings of Gritsch [48]. The skewed coolant flow injected from the diffuser tends to separate from one of the sidewalls in the diffuser section leading to reduced lateral spreading in the near hole region. This effect is mainly dominated by the relatively large lateral expansion angle and can be reduced if the diffuser geometry is changed to smaller expansion angles.

Comparing the centerline and laterally averaged film effectiveness for the investigated blowing ratios as done in Fig. 7.7, it can be stated that film injection through the diffuser holes shows no remarkable influence of the density ratio. For the test case with a blowing ratio of $M = 0.50$, the curves obtained for an elevated density ratio of 1.37 using CO_2 and for air injection with a density ratio of 0.90 are nearly the same. This is a direct result of the diffuser effect on the injected coolant reducing the jet momentum. Improving the lateral spreading of the coolant further leads to smaller penetration depths of the film-cooling jets into the main flow.

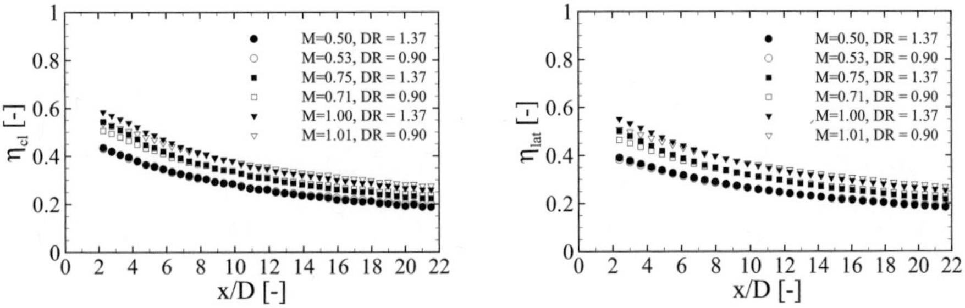

Figure 7.7: Centerline η_{cl} and laterally averaged effectiveness η_{lat} obtained for laid-back fan-shaped holes with $M = 0.5$, 0.75 and 1.0 using CO_2 and air injection, $Re_{m,D} = 7700$, $u_m = 22.9\,m/s$, $Tu = 1.5\,\%$

Slightly higher film effectiveness is seen for the high density ratio case and blowing rates of $M = 0.75$ and 1.00 which occurs closely downstream of the injection location. This may be attributed to the flow conditions inside the diffuser sections with the higher momentum flux ratio for air injection causing local flow separation on the downstream diffuser surface. This results into enhanced jet detachment from the wall downstream of the hole exit. Hence, the increased film effectiveness, that was found in the high density ratio cases, indicate improved attachment of the film layer to the cooled surface. This effect significantly diminishes with larger streamwise distances. It coincides with the findings obtained from the spatial film-effectiveness distributions.

While the laterally averaged film effectiveness shows essentially the same level as the centerline data, the effect of the density ratio is slightly less pronounced. Thus, film injection through diffuser holes gives significantly higher laterally-averaged film effectiveness compared to cylindrical holes due to their discrete jet nature. The diffusor part of the hole forces the flow velocity and jet momentum to decrease. It further enhances the lateral spreading of the film jets which leads to a broader lateral distribution of the coolant and improved wall coverage downstream of the holes. Adjacent film jets are merged early at their edges as the coolant proceeds downstream which mainly depends on the chosen pitch-to-diameter ratio. Therefore, it forms a closed coolant layer without any gaps allowing for hot gas ingestion.

7.5 Comparison to Literature Results

Prior to its application of the TSP-method in endwall cooling investigations, the flat plate data for the diffuser holes is compared to results from previous studies. Therefore, three different test cases were selected (see Table 7.3) wherein all authors established film injection through fan-shaped holes. While the geometrical parameters of the holes are identical in every detail except for the hole diameter between the present study and the work done by Saumweber et al. [112] and Gritsch et al. [48], differences are seen in the lateral and axial diffuser angle as well as in the hole pitch-to-diameter ratio for the hole geometry used by Brauckmann [16]. It should also be mentioned that Gritsch et al. [48] conducted film effectiveness measurements on a single cylindrical and a shaped hole while Saumweber et al. [112] used a row of three scaled-up film-cooling holes. The pitch-to-diameter ratio of the present study and the experiments done by Saumweber et al. [112] with P_h = 4 differs from the ratio used in Brauckmann [16] with P_h = 6.

(a)	L_h/D [-]	P_h/D [-]	α [°]	γ_{lat} [°]	γ_{ax} [°]	D [mm]	N [-]
present study	6	4	30	14	15	5	5
Brauckmann [16]	6	6	30	10	10	2.5	5
Gritsch et al. [48]	6	-	30	14	15	10	1
Saumweber et al. [112]	6	4	30	14	15	5	3

(b)	M [-]	D_R [-]	I [-]	Tu [%]	Ma_m [-]	$Re_{m,D}$ [-]	δ_{99}/D [-]
present study	0.5-2.0	1.37	0.19-2.92	1.5	0.066	7701	3
Brauckmann [16]	0.5-2.0	1.84	0.14-2.17	1	0.3,0.6	-	-
Gritsch et al. [48]	0.25-2	1.85	0.033-2.16	< 2	0.6	$1.3 \cdot 10^5$	0.5
Saumweber et al. [112]	0.50-2.5	1.70	0.15-3.68	3.6	0.3	$0.25 \cdot 10^5$	-

Table 7.3: Comparison of geometrical (a) and experimental boundary conditions for diffuser hole investigations (b)

Also, distinct differences can be seen for the main-flow conditions. While the present study is conducted at very low main flow velocities leading to Mach numbers of 0.066, the experimental data from the literature is obtained altogether at higher Mach numbers of 0.3 and 0.6 respectively. The differing experimental boundary conditions in combination with the varying hole diameter lead furthermore to much higher coolant Reynolds-numbers in the previous studies. Additionally, all authors conducted the film-cooling measurements at very high temperature levels with the main flow heated over 500°C and the coolant injected at ambient conditions.

This leads to density ratios of about 1.8 which is much higher than what could be established in the present experiments with CO_2-injection which equals only $D_R = 1.37$.

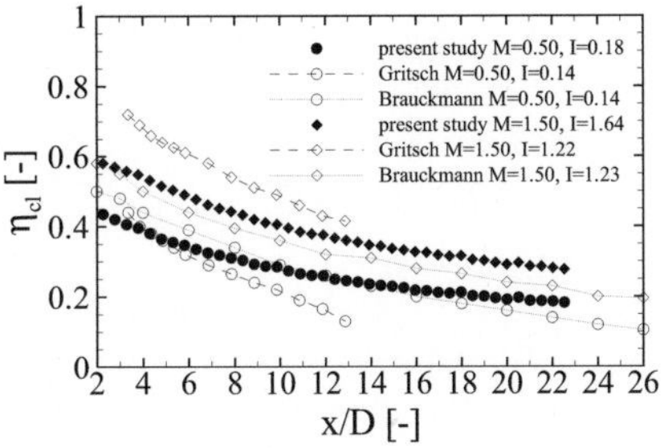

Figure 7.8: Comparison of centerline film effectiveness to previous literature results for two blowing rates $M = 0.5$, $I = 0.19$ and $M = 1.50$, $I = 1.64$ using CO_2-injection, $D_R = 1.37$, $Re_{m,D} = 7700$, $u_m = 22.9\,m/s$, $Tu = 1.5\,\%$

Comparison for the film effectiveness along the jet centerline is shown in Fig. 7.8 which displays the data for two selected blowing cases. The first one has basically a moderate injection strength at $M = 0.50$ and the other has an increased blowing rate of $M = 1.50$. Considering the test case with a moderate blowing rate of $M = 0.50$, the highest film effectiveness of 0.45 - 0.5 is observed downstream of the hole exit and shows overall good agreement between the compared data sets. With increased streamwise distance, the decline in film effectiveness, as observed by Gritsch et al. [48], is more pronounced than in the present study. This may be attributed to the exposed position in the hot-gas flow path that is produced by a single film-cooling hole. Due to that, the main flow is more easily be able to penetrate the film layer. The jets face stronger interaction and mixing at their boundaries which leads to enhanced decrease in film effectiveness in downstream direction. Smaller differences are observed in comparison with Brauckmann [16] where slightly lower film effectiveness is found near the hole exits. The decay of film effectiveness in streamwise direction is less pronounced in the present case which leads to slightly higher film effectiveness beyond streamwise distances of $x/D > 16$.

As the blowing rate is increased to 1.50, the comparison of the centerline film effectiveness with the data obtained by Brauckmann [16] indicates a qualitatively similar trend in streamwise film effectiveness decay. Although its magnitude is constantly higher about ~ 0.04 for the present study, it seems that the streamwise behavior of the film jets, yet the physical process of coolant interaction and mixing may be similar in both cases.

The comparison of the results is continued also for the laterally-averaged film effectiveness, as shown in Fig. 7.9. For the test case with a low blowing rate of $M = 0.50$, quite good agreement of the film effectiveness close to the hole exit is found between the present study and the data from Saumweber et al. [112]. The development in streamwise direction shows a similar decay but the laterally-averaged film effectiveness of the present study is slightly higher of about 0.04. This may be due to lateral heat conduction effects induced by the hot film layer. At small blowing rates, the lateral spreading of the film is less intense. This leads to enlarged, uncooled wall regions between the jets which act as heat source. Due to the temperature gradient between the hot jet core and the uncooled wall regions, the wall temperature is affected and results into elevated laterally-averaged film effectiveness.

Larger differences are seen when the laterally-averaged film effectiveness is compared to the data from Gritsch et al. [48] and Brauckmann [16]. In these cases, the differences are either

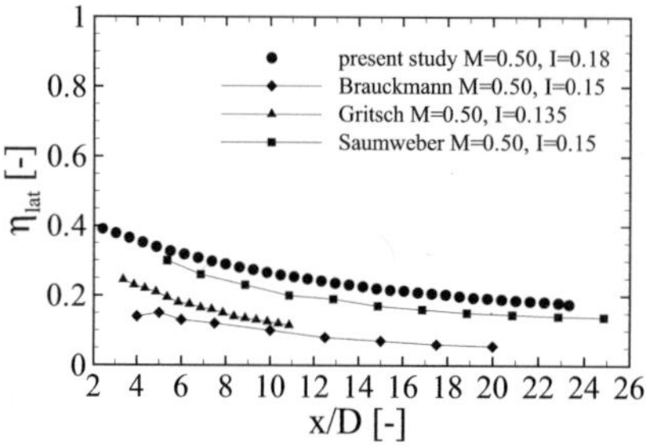

(a) $M = 0.50$, $I = 0.19$

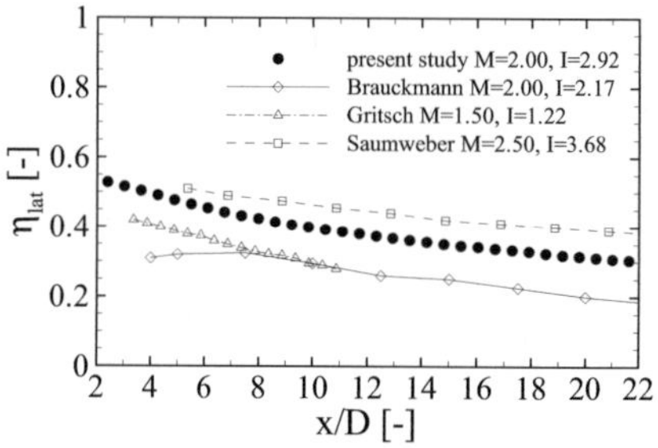

(b) $M = 2.00$, $I = 2.92$

Figure 7.9: Comparison of laterally-averaged film effectiveness for the shaped holes to previous literature results at two different blowing rates using CO_2-injection, $D_R = 1.37$, $Re_{m,D} = 7700$, $u_m = 22.9\,m/s$, $Tu = 1.5\,\%$

attributed to the use of a single film hole as done by Gritsch et al. [48] or the larger film hole pitch ratio for the holes used by Brauckmann [16]. This results into a smaller coverage value of $Cov = t_h/P_h = 0.522$ compared to the holes used in the present work with $Cov = 0.933$. The cooling film released from a single fan-shaped film hole is strongly opposed to the hot gas at the film layer's edges. It leads to an enhanced mixing between the coolant and the main flow in this region, hence to a faster decay of the film effectiveness in lateral direction. Similarly, the increase of the hole pitch leada to an increased lateral distance between two cooling films which in turn results into reduced wall coverage and decreased laterally-averaged film effectiveness.

It is remarkable that for the high blowing rate of $M = 2.00$ the measurements done by Gritsch et al. [48] and Brauckmann [16] show quite good agreement in the far field of the holes ($x/D > 6$) despite the different hole geometries in terms of number of the holes and diffuser opening angles. Closer to the hole injection ($2 < x/D < 6$), the data from Gritsch et al. [48] exhibits higher film effectiveness compared to Brauckmann [16] which occurs coincidently for both blowing ratios. This effect may be attributed to the larger lateral expansion angle γ_{lat} used by Gritsch et al. [48] which produces enhanced lateral spread of the coolant and therefore improved coverage. This effect is strictly limited to the near-field of the hole injection location and diminishs with increasing streamwise distance. These findings suggest that the film row with a large pitch-to-diameter ratio of 6 and moderate opening angle of 10°, as used by Brauckmann [16], delivers an equivalent cooling performance compared to a single diffuser hole with large lateral opening at 14°. Comparing the two previous cases to the present study for the high blowing ratio, the laterally averaged film effectiveness in the latter is constantly higher about 0.11 but shows comparable qualitative trends for the streamwise decay beyond $x/D > 8$.

Comparing with Saumweber et al. [112], slightly lower laterally-averaged film effectiveness about 0.07 is observed which is mainly attributed to the lower blowing rate of the present work. The decay of the film effectiveness in streamwise direction is qualitatively similar for both data sets which indicates comparable physical behavior of the film jets. However, the constantly higher film-effectiveness level which was observed from Saumweber et al. [112] would be reduced if the data is extrapolated to the hole exit.

8 Endwall Film-Cooling

8.1 Cascade Aerodynamic Performance

8.1.1 Loss Characteristic and Secondary Loss Coefficient

Characteristic values as the profile loss ζ_p and the outlet flow angle β_2 are plotted in Fig. 8.1 versus the entire range of inlet flow angles β_1 that can be achieved by turning the cascade. The profile-loss characteristic shows a typical pattern for a turbine cascade at airfoil midspan. Small changes in profile loss can be observed over a broad range of inlet flow angles.

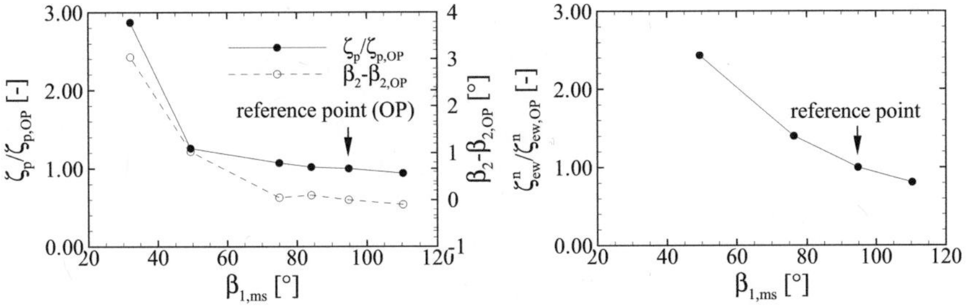

Figure 8.1: Normalized loss characteristic of airfoil, normalized net endwall loss coefficient $\zeta_{ew}/\zeta_{ew,OP}$ and outlet flow angle versus inlet flow angle $\beta_{1,ms}$ at midspan

The design point is determined at an inlet flow angle of $\beta_{1,ms} = 94.8°$ corresponding to a profile loss of 2.84 %. A small change in profile loss of less 5 % is found within a region of ± 10° incidence. This strong insensitivity is a direct result of the large leading edge radius. The smallest profile loss according to the measurement results is observed at $\beta_{1,ms} = 110°$ which equals a decrease in cascade turning of 20° with regard to the operating point. Increasing incidence leads to a slow increase in profile loss and cascade turning. At an inlet flow angle of 49.4°, the profile loss is 3.6 % which equals about 27 % rise with regard to the design point.

At an inlet flow angle $\beta_{1,ms} = 30°$, the profile loss starts into a sudden rise up to 8 %. This indicates boundary-layer separation around the profile. These findings coincide with a significant drop of the measured outlet flow angle of about 3° indicating distinct loss in the cascade's turning capacity. The isolated secondary loss coefficient is enhanced by more than a factor of two at almost 40° incidence.

Based on these conclusions, the endwall cooling experiments are carried out for three different inlet flow angles as given in Table 8.1. Three loading cases were specified with the reference test case near zero incidence ($\beta_{1,ms} = 94.8°$), minimum loading) which is close to the nominal operating point of the real-engine vane. The other two test cases were done at an incidence of +15° ($\beta_{1,ms} = 75°$, moderate loading) and +40.6° ($\beta_{1,ms} = 49.4°$) which is considered the case at maximum loading.

8.1.2 Inlet Boundary Layer

Prior to the investigations of the cascade exit flow, the inlet-boundary layer and the turbulence intensity were measured using a pitot and single hot-wire probe. The probes were inserted at a location of x/L_{ax} = -1.52 upstream of the trailing edge. A summary of the corresponding boundary-layer parameters of the incoming cascade flow can be found in Table 8.1 for the investigated loading cases. For the experimental test cases reported here, higher airfoil

Table 8.1: Parameter of the inlet boundary layer

β_{cas} [°]	$\beta_{1,ms}$ [°]	i [°]	δ^* [mm]	θ [mm]	H [-]	$\overline{\overline{\zeta}}_{1,ibl}$ [-]	Ma_1 [-]	Re_1 [-]
90	94.8	-4.8	4.655	3.948	1.179	0.005690	0.05	$2.6 \cdot 10^5$
70	75.0	+15	4.695	3.985	1.178	0.008387	0.06	$2.8 \cdot 10^5$
50	49.4	+40.6	4.558	3.876	1.176	0.011158	0.08	$3.3 \cdot 10^5$

loading was achieved by increasing the incidence of the inlet flow by turning the cascade anticlockwise. Therefore, the cascade angle β_{cas} relates the position of the cascade assembly to the circumferential direction. Only small changes in the boundary-layer parameters in terms of the displacement, momentum thickness and shape factor are found for increasing incidence. The area-averaged total pressure loss coefficient $\overline{\overline{\zeta}}_{1,ibl}$ is increased up to 49 % for the test case at maximum loading.

The inlet Mach number is calculated to 0.05 for the reference case which results into an inlet flow angle of $\beta_{1,ms} = 94.8°$ at airfoil midspan. The Mach number is slightly increased as the incidence increases due to acceleration of the incoming flow. It's caused by the reduction of the inlet cross-sectional area when the upper and lower tunnel walls are lowered to match the airfoil cascade. This is done to avoids air leakage through the gaps between the airfoils and the connecting tunnel boards. The corresponding inlet Reynolds number Re_1 determined with the inlet flow condition and the profile length is also given in Table 8.1.

The potential flow effect of the airfoil on the incoming flow is obvious in the diagram for the

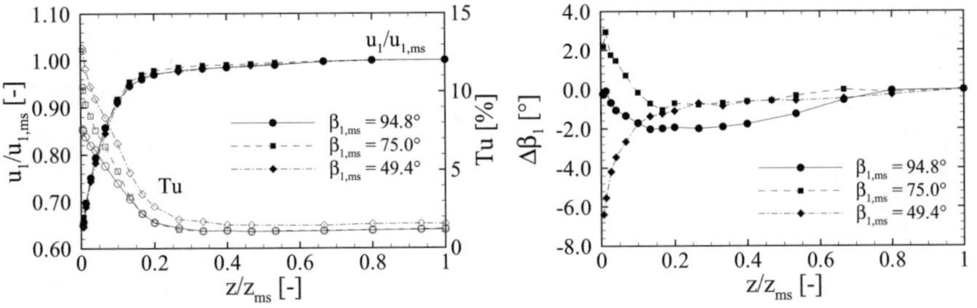

Figure 8.2: Normalized velocity profile at cascade inlet $u_1/u_{1,ms}$ and flow angle deviation $\Delta\beta_1$ along spanwise direction (MP1, x/L_{ax} = -1.52) with variation of loading

flow angle deviation $\Delta\beta_1$. The curves show a distinct spanwise region of varying flow angles which is located near the endwall up to 40 % midspan. The largest deviation of the spatial flow angle is found for the test case at maximum incidence yielding $\Delta\beta_1$ of -6°. It results into less flow turning and decreased loading of the profile in that spanwise area. Less severe changes in the spanwise flow angles of $\pm 2°$ are found for the other test cases. At airfoil midspan, low turbulence intensity of 1.2 % is observed being fairly constant over a large spanwise section of 75 %. At a wall distance of z/z_{ms} = 0.25 where the boundary layer begins to grow, it rapidly increases up to values of 7.4% close to the wall. The turbulence intensity near the endwall rises with increased loading which yields about 12.3 % at maximum loading.

8.1.3 Flow Field at Cascade Exit

The spatial total pressure-loss field at the cascade outlet is displayed in Fig. 8.3 for the three incidence conditions. Concentration of boundary-layer material on the suction side of the airfoil is clearly visible in all cases which is then shifted toward the airfoil midspan with higher incidence due to the enhanced secondary flow. The pattern of the measured total pressure losses shows a so-called "dog-bone" geometry typical for highly-loaded turbine cascades. It is characterized by a twin-peak pattern with the less pronounced peak positioned itself close to the core of the channel vortex. The second loss peak shows larger loss gradients and represents the discrete trailing edge vortex. It develops as a discrete vortex from the furled trailing edge vortex sheet and collects the low energy material from the profile boundary layer around its core. At the vane's operating point, both loss peaks partially merge with each other developing a single zone of high losses.

The merged loss peak is observed at y/P_p = -0.70 and z/z_{ms} = 0.27 for reference case at $\beta_{1,ms}$ = 94.8° yielding a total pressure loss coefficient of ζ_2 = 0.234. The maximum spatial loss coefficient is found to be close to the endwall at a pitchwise location of y/P_p = -1.1. The

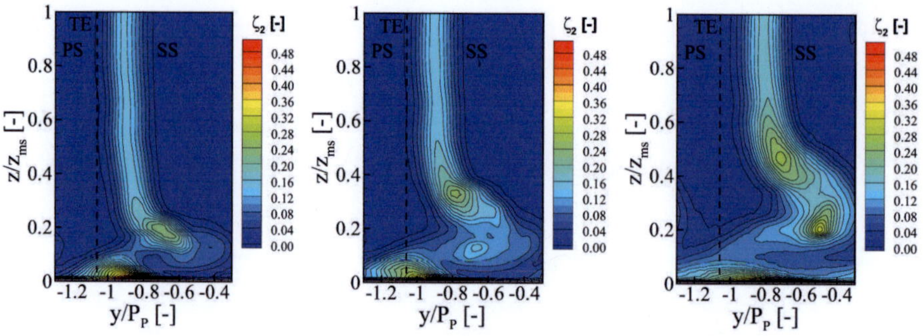

(a) $\beta_{1,ms} = 94.8°$ (b) $\beta_{1,ms} = 75.0°$ (c) $\beta_{1,ms} = 49.4°$

Figure 8.3: Total pressure loss coefficient ζ_2 at cascade exit (MP4, $s/L_p = 1.03$) with variation of loading

wake at midspan is found to be purely two-dimensional. The profile loss for the operating point was determined to 2.84%, increasing with higher airfoil loading up to 3.59%. The extent in pitchwise direction is about 2% of the airfoil pitch.

With decreased inlet flow angles, hence higher incidence the endwall loss region grows in spanwise and pitchwise direction which is mainly attributed to the enhanced vortical motions of the secondary flow. The loss peaks associated with either the trailing edge vortex or the passage vortex can be observed as seperate loss zones. At minimum loading, the extent of the endwall loss region in spanwise direction is estimated to approx. 40% of blade midspan. With increased incidence to maximum loading, the endwall loss region is significantly enlarged, measuring a spanwise extent of $z/z_{ms} = 0.75$. Both loss peaks show larger loss gradients which contributes to the enhanced secondary flow structure. The extent and depth of the trailing edge wake as well as the averaged profile loss coefficient increases with higher incidence.

The local distribution of the outlet flow angle for all incidences (Fig. 8.4) shows regions of distinct under- and overturning. For the vane's operating point at $\beta_{1,ms} = 94.8°$ the maximum overturning is determined to -10.8° located close to the endwall at $z/z_{ms} = 0.0095$ and the underturning with its maximum of +5° located $z/z_{ms} = 0.130$ which equals 6% blade span. These distinct regions of underturning and overturning of the outlet flow angle characterize the movement of the passage vortex as directed from the pressure toward the suction side near the endwall. As the fluid is transferred across the passage towards the SS it causes larger flow deflection in this spanwise region. When the crossflow approaches the airfoil suction side it disturbs the newly growing wall boundary layer while it is deflected toward the midspan section. The movement is then further continued back toward the pressure side forming the passage vortex which then leads to less flow turning in that region. The maximum overturning

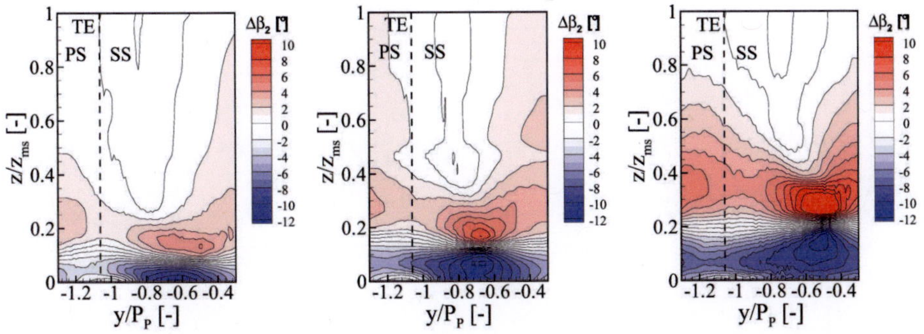

(a) $\beta_{1,ms} = 94.8°$ (b) $\beta_{1,ms} = 75.0°$ (c) $\beta_{1,ms} = 49.4°$

Figure 8.4: Local flow angle deviation $\Delta\beta_2$ at cascade exit (MP4, $s/L_p = 1.03$) with variation of loading

occurs in the region of the highest losses. This can be related to the location where the passage vortex changes its direction toward the pressure side. The experimental results for the highest cascade loading at $\beta_{1,ms} = 49.4°$ show more pronounced under- and overturning. The maximal overturning is determined to -11.2° at $z/z_{ms} = 0.278$ which equals 7% blade height and the maximal underturning of max. +14° at $z/z_{ms} = 0.28$ (13% blade height). For the test case at high incidence, the minimum and maximum of the outlet flow angle occurs at a larger wall distance of $z/z_{ms} = 0.37$.

These findings coincide with the pitchwise averaged results depicted in Fig. 8.5 for the endwall loss coefficient ζ_{ew} and outlet flow angle deviation $\Delta\beta_2 = \beta_2 - \beta_{2,ms}$. Thereby, positive values of $\Delta\beta_2$ imply underturning of the cascade flow and negative values a flow overturning. Observing the results of the endwall loss coefficient for the reference case, the growth of ζ_{ew} starts at 25 % midspan. It steadily increases with decreasing spanwise position to values of $\zeta_{ew} = 0.15$ close to the endwall. With decreasing inlet flow angles, the endwall loss region extends in spanwise direction and starts earlier at larger distances from the endwall. For the test case with large incidence ($\beta_{1,ms} = 49.8°$) the position is observed to be shifted to 65 % midspan. The loss peaks are also clearly visible in the spanwise distribution which show larger extent and depth with increased incidence.

The vortex motion and its impact on the cascade outlet flow is more obvious in the plot for $\Delta\beta_2$ where the curves show distinct regions of decreased and increased deflection. It is related to the flow distortion caused by the movement of the passage vortex from the PS toward the SS of the adjacent vane which is continued back toward the PS far from the endwall. The impact of the corner vortex is seen for spanwise coordinates z/z_{ms} less than 5 percent. It weakens the local deflection enhancement caused by the channel vortex close to the endwall.

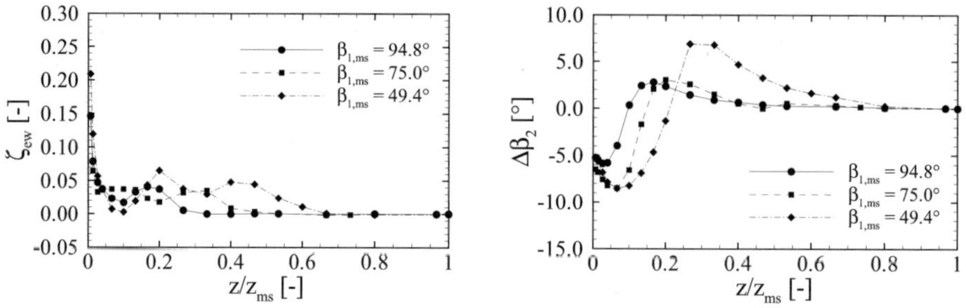

Figure 8.5: Pitchwise-averaged endwall loss coefficient ζ_{ew} and flow angle deviation $\Delta\beta_2$ versus the spanwise direction (MP4, s/L_p =1.0) with variation of loading

This vortex develops in the corner region between the suction side and the endwall and rotates anti-clockwise compared to the channel vortex.

The effects are found to be enhanced with higher incidence due to the intensified secondary flow's motion. The spanwise locations of underturning and overturning are shifted toward the airfoil midspan. A local maximum and minimum of the flow angle deviation appear at different spanwise locations. For the reference case, the maximum of overturning of -6° is found at z/z_{ms} = 0.03 whereas the maximum of underturning of +3° can be found at larger endwall distances of z/z_{ms} = 0.16. The distinct regions of over- and underturning are enhanced with higher cascade incidence. The spanwise location of the pitchwise peaks is found to be shifted toward the airfoil midspan. Furthermore, this effect is also enhanced with higher incidence.

Additional measurements have been undertaken to investigate the strength of the secondary flow downstream of the cascade and to visualize the vortex structures. Therefore, five-hole probe traverses are carried out at similar streamwise distances and loading conditions as the pitot-probe measurements. The flow field including the circumferential and radial flow angle is determined at normalized distances downstream of the cascade of s/L_p = 0.34, 0.70 and 1.03. The summary of all experimental results for the five-hole probe traverses can be found in the appendix A.5.1. Subsequently, only the results on measurement plane s/L_p = 1.03 are discussed. The secondary flow vectors and the distribution of the streamwise vorticity is plotted in Fig. 8.6 with variation of loading.

At the vane's operating point ($\beta_{1,ms}$ = 94.8°), intense vortical movement was observed with its center at y/P_P = -1.0 near the endwall at 10 % wall distance. Two vortices can be distinguished with the passage vortex (PV) dominating the near-wall region at the cascade outlet due to its largest strength of rotation. Also the trailing edge vortex (TV) with its opposite rotational sense can be observed, placed farther off the endwall at app. 40 % wall distance. The flow structure of the small corner vortex is hardly recognized and can only be identified by an

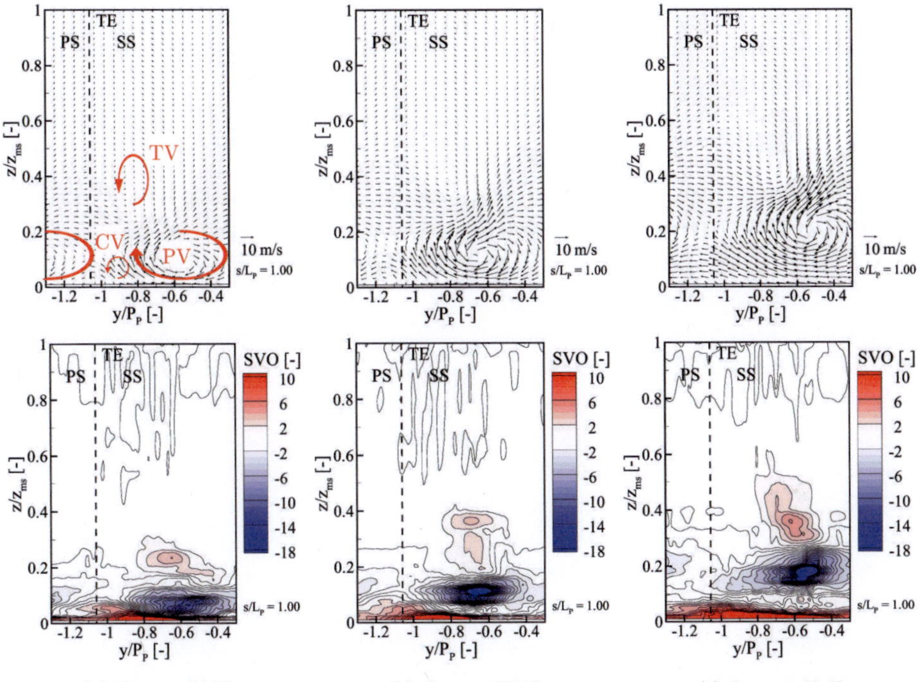

(a) $\beta_{1,ms} = 94.8°$ (b) $\beta_{1,ms} = 75.0°$ (c) $\beta_{1,ms} = 49.4°$

Figure 8.6: Secondary flow vectors and streamwise vorticity SVO at cascade exit (MP4, s/L_p =1.03) with variation of loading a) $\beta_{1,ms} = 94.8°$, b) $\beta_{1,ms} = 75.0°$ and c) $\beta_{1,ms} = 49.4°$

enhanced cross flow. With increased loading, the vortex center shift away from the endwall about 12 %, increasing the pitchwise and spanwise extent of the vortical motion. Intensified vortical movement with higher loading is identified by increased magnitudes of the secondary flow velocities. The streamwise behavior of the secondary flow's vortex pattern is displayed in Fig. A.10 - Fig. A.12 in the appendix.

Along the flow path, the secondary flow vectors clearly indicate an enhanced vortex spreading in pitchwise and spanwise direction. Also, the center of rotation moves away from the wall and slightly in lateral direction when proceeding downstream. Due to diffusion processes, the rotational motion of the vortex is slightly weakened. This further coincides with enhanced lateral and spanwise loss regions at greater streamwise distances. The maximum in pressure loss close to the endwall significantly increases in extent and amount. This is mainly due to the alternation and re-distribution of the endwall boundary layer along with increased friction losses. Larger lateral extent of the midspan wake is found in streamwise direction due to mixing within the shear layers of wake and free-stream. Likewise, the depth of the wake is significantly reduced. With increased incidence, the shape of the passage vortex is slightly deformed, showing enhanced secondary flow vectors near the suction side, indicating intensified upward flow movement. Similar effects are seen as closer the measurement plane is placed to the cascade outlet.

When the streamwise vorticity SVO is examined, two distinct regions with elevated streamwise vorticity are identified. The first one is located closer to the endwall and corresponds quite well with the position of the passage vortex's center. Hence, this peak may be associated with the vortical movement of the passage vortex. It's peak vorticity is obtained to $SVO = -12$ with its position at y/L_P = -0.53 and 7 % spanwise distance with respect to profile midspan. The other region is located farther away from the endwall showing a less intense and positive peak of the SVO at 4. It may be attributed to the weaker trailing edge vortex having an opposite sense of rotation. The first SVO-peak, associated with the passage vortex, is increased by the factor three at maximum cascade loading. This is mainly attributed to the more pronounced vortical strength of the secondary flow with higher loading. The SVO-peak of the trailing-edge vortex remains approximately constant for $s/L_P = 1.03$. Only a small increase of the vorticity to $SVO = 5$ is observed. Since the vortical motion is reduced along the flow path, the SVO-peak of the trailing-edge vortex does not show any deviations at this streamwise position with varying loading conditions. Nevertheless, a visible enhancement of the trailing-edge vortex can be observed by its more pronounced SVO-peak at the shortest streamwise distance $s/L_P = 0.34$, as displayed in the appendix A.5.2).

8.1.4 Effect of Flow-Path Distance and Airfoil Fillet

Recent studies on secondary losses in turbine cascades done by Sauer et al. [111] have not only shown that endwall losses grow considerably with increasing streamwise distance downstream of the cascade exit but are also influenced by the chord length of the profile. The authors found that the endwall total-pressure loss is directly proportional to the chord length for a measurement distance $s/L_P = 1.0$. The loss growth in streamwise direction is mainly attributed to the re-distribution of the endwall boundary layer in combination with the alternation process of the secondary flow. The investigations further provide a large database with wide parameter variation of the velocity ratio and chord length. Therefore, the endwall pressure losses that are obtained at three different measurement distances ranging from $s/L_P = 0.39$, 0.70 to 1.03 are compared to the data from Sauer et al. [111], as displayed in Fig. 8.7. The secondary loss coefficient is plotted for the used pitch-to-chord length ratio of 0.968 and the velocity ratio $v = 3.5$. The approach to derive this parameter is outlined in chapter 5.1 and is identical to the methodology applied by Sauer et al. [111].

Firstly, enhanced endwall losses are observed with increasing distance in streamwise direction with its magnitude significantly increased at higher loading conditions. Lowest endwall losses

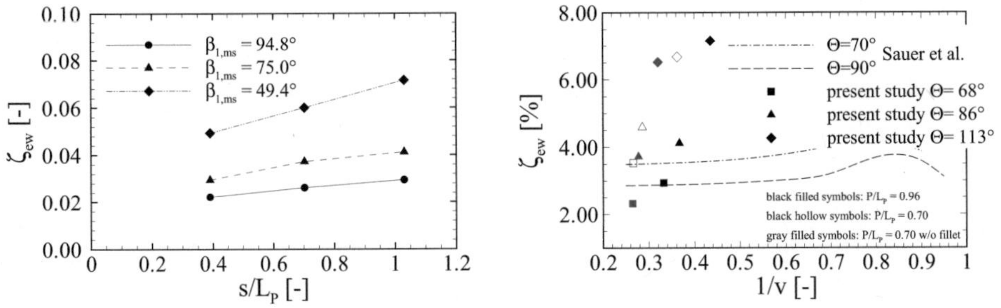

(a) influence of streamwise distance

(b) influence of profile pitch and fillet

Figure 8.7: Endwall loss coefficient ζ_{ew} versus streamwise distance s/L_P with variation of loading (a) and comparison of present endwall loss coefficients for two pitch-to-chord length ratios of $P_P/L_P = 0.7$ and 0.968 with experimental data from Sauer et al. [111] obtained for profiles for $P_P/L_P = 0.7$ at a streamwise distance of s/L_P =1.0 (b)

of 2.2 % can be found for the smallest measurement distance at $s/L_P = 0.34$ and the reference case which then rises up to 4.9% as cascade turning, hence the airfoil loading increases to incidences of 40°. At greater streamwise distances, the endwall losses grow remarkably about 33 % considering the reference test case at $\beta_{1,ms} = 94.8°$. It is furthermore visible that the growth of the endwall losses is enhanced with higher cascade turning leading to larger magnitudes of about 45 % for $s/L_P = 1.03$ and the test case of $\beta_{1,ms} = 49.8°$.

Comparing the results to Sauer et al. [111], quite good agreement can be stated for their test case with the smallest cascade turning of $\Theta = 70°$ which equals to the present case with an inlet flow angle of $\beta_{1,ms} = 94.8°$. As the airfoil turning increases to $\Theta = 90°$, a difference of 0.5 percentage points occurs in the measured secondary loss. The deviations become more pronounced at even higher cascade turning which indicates a different process of loss generation and development of the secondary flow between the two cascade setups.

In the present work, the used larger pitch-to-chord length ratio attributes to the enhancement of the secondary losses. It enlarges the wetted endwall area covered by the cross flow which results into increased friction losses. Moreover, the fillet at the airfoil root, which transfers the cylindrical vane sleeve to the endwall, affects the development of the passage vortex. It promotes its downward movement on the pressure and the upward movement of the near-wall cross flow on the suction side. Furthermore, the fillet with its wall surface introduces additional friction losses and changes the pressure field in that region, such that the development of corner vortex is suppressed. To separate the effects from each other, first the pitch-to-chord length of the used vane is reduced to 0.70 and afterwards the fillet is removed from the two center vanes providing a straight, cylindrical profile. The total pressure loss measurements were repeated for the three cascade turning cases but only at the streamwise distance of $s/L_P = 1.03$.

For the reduced pitch-to-chord length ratio, the secondary loss coefficient at a cascade turning of $\Theta = 70°$ is significantly increased about 20 % to $\zeta_{ew} = 0.03535$ which interestingly shows a good comparability with the curve of 90° deflection given by Sauer [111]. With further increase of loading, the endwall loss coefficient tends to give a slightly lower increase with higher cascade turning but shows the same trend as observed for the original pitch-to chord length ratio of 0.967. The highest incidence delivers an endwall loss coefficient of $\zeta_{ew} = 0.066753$ which is about 7 % lower than the test case with the large airfoil pitch ratio.

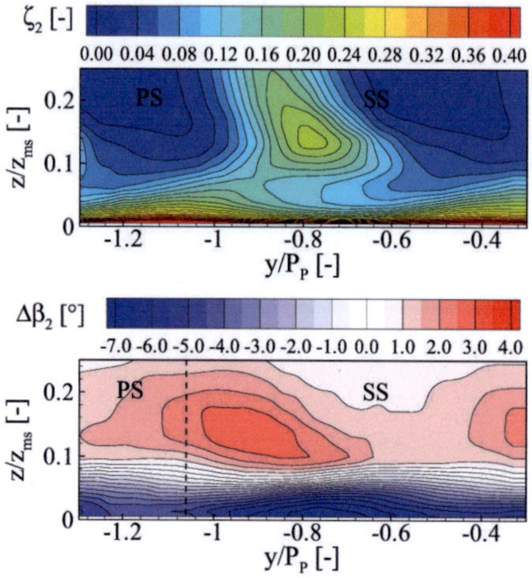

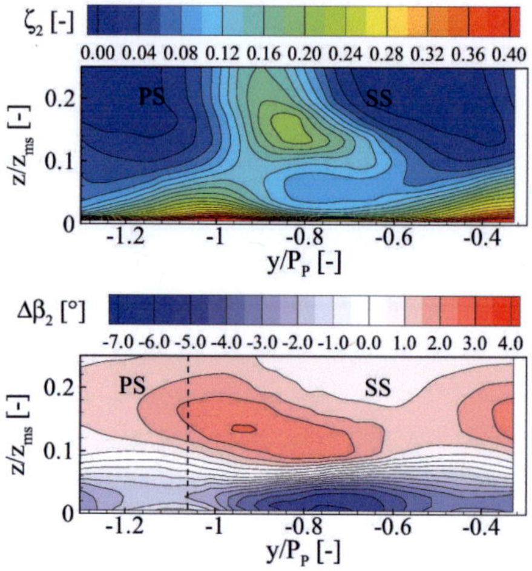

Figure 8.8: Effect of fillet geometry on local loss distribution and flow angle deviation, $\beta_{1,ms} = 94.8°$, $P/L_P = 0.7$, $s/L_P = 1.03$

With removing the fillet, the secondary loss coefficient shows a remarkable drop of about 35 % down to 0.02318. This lies only about 0.6 % points of loss coefficient beneath the curve from Sauer et al. [111] obtained for $\Theta = 70°$. When the incidence is increased to reproduce the test case of $\Theta = 90°$, smaller reduction of the endwall losses was obtained. Without fillet, the loss coefficient still shows an almost 20 % drop compared to test case with the fillet. Here, the obtained endwall loss of 0.0372 agrees quite well with the data from Sauer et al. [111]. Further increase of the cascade incidence up to 40° (cascade turning $\Theta = 116°$) leads to smaller reduction in endwall loss when the fillet is removed. This yields a secondary loss coefficient of $\zeta_{ew} = 0.06520$ which equals a reduction of less than 3 % compared to the results with vane fillet.

The profile fillet tremendously change the development of the secondary flow vortices along with the related total pressure losses. While the effect on the endwall losses is significant for zero up to moderate incidences of 20°, it diminishes almost completely with higher incidences up to 40°. The results suggest that the fillet's enhancing effect on the down- as well as upward flow movement of the passage vortex along the airfoil surfaces is balanced out by the enhanced pitchwise pressure gradient as driving potential for the secondary flow.

To get a deeper insight on the flow physics with the fillet involved, the distribution of the total-pressure loss coefficient and the flow-angle deviation is shown in Fig. 8.8. The results are plotted for the small pitch-to-chord length ratio of P_P/L_P = 0.7 at a streamwise distance of s/L_P = 1.03. For the test case with fillet at an inlet flow angle of $\beta_{1,ms}$ = 94.8°, the total pressure loss coefficient shows a constant high level of magnitude for the spanwise sections closest to the endwall. In the pitchwise distribution, no concise maximum loss peak is formed which is typical for cylindrical airfoils in turbine cascades. When the fillet is removed, the total pressure loss changes to reveal a distinct loss peak near the endwall and rapidly decreases further distant from the maximum loss at y/P_P = -1.

Furthermore, the plots for the flow-angle deviation show a visible influence when the fillet is removed by reducing the depth and extent of the regions with increased under- and overturning. The maximal overturning with fillet is obtained with $\Delta\beta$ = -6.6° located close to the endwall at y/P_P = -0.75. It is reduced to $\Delta\beta$ = -5° with removed fillet which is attributed less pronounced overturning. The influence of the fillet on the maximal underturning is found to be less pronounced and located at larger wall distances. It results into a decreased flow-angle deviation of about 0.6° with removal of the fillet.

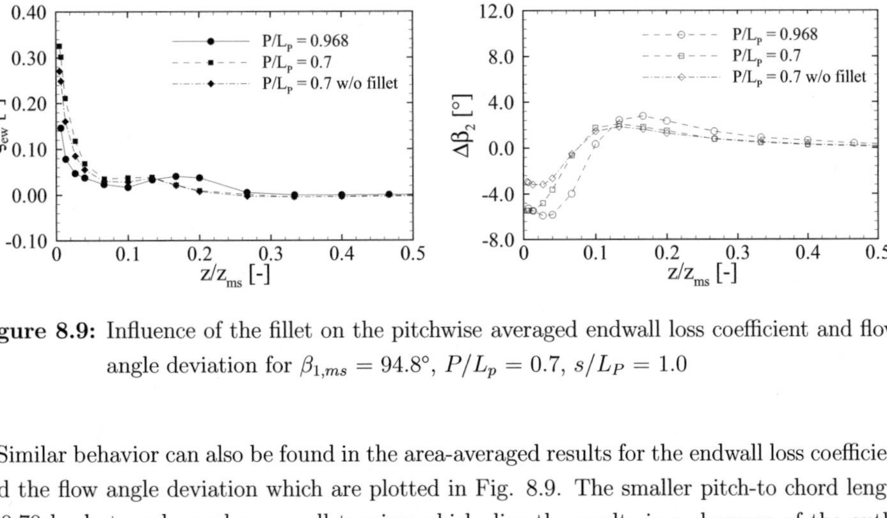

Figure 8.9: Influence of the fillet on the pitchwise averaged endwall loss coefficient and flow-angle deviation for $\beta_{1,ms}$ = 94.8°, P/L_p = 0.7, s/L_P = 1.0

Similar behavior can also be found in the area-averaged results for the endwall loss coefficient and the flow angle deviation which are plotted in Fig. 8.9. The smaller pitch-to chord length of 0.70 leads to enhanced near-wall turning which directly results in a decrease of the outlet flow angle of about 6.7°. With reduction of the profile pitch, the entire vane passage with its throat is reduced. The pressure and suction side of the adjacent airfoils draw closer to each other. Hence, the measured profile loss at airfoil midspan rises about the factor of almost two showing values of $\zeta_P = 0.053$ for an inlet flow angle of $\beta_{1,ms} = 94.8°$. Also the extent and depth of the profile wake is increased while it partly merges with the adjacent wakes, as displayed in Fig. 8.10. This leads to elevated total pressure losses in the free-stream region. Integration over the pitchwise direction then results into increased profile losses.

Comparing the endwall losses for both pitch-to-chord ratios, the spanwise evolution shows a significant increase toward the endwall. Here, the loss maximum is $\zeta_{ew} = 0.30$ for the small pitch ratio. This equals twice the endwall pressure loss obtained for the original pitch ratio of nearly unity. Removing the fillet results into a decrease of the loss maximum down to $\zeta_{ew} = 0.27$ which equals a 16 % drop. The spanwise evolution of the endwall loss coefficient shows only small differences for the test cases with and without fillet and the small pitch ratio of 0.7. Comparing this to the results of the large pitch ratio, the loss peak is observed at larger wall distance of $z/z_{ms} = 0.17$. With reduced airfoil pitch, the loss peak is found to be closer to the endwall at $z/z_{ms} = 0.125$ and slightly less pronounced. It's position does not change when the fillet was removed.

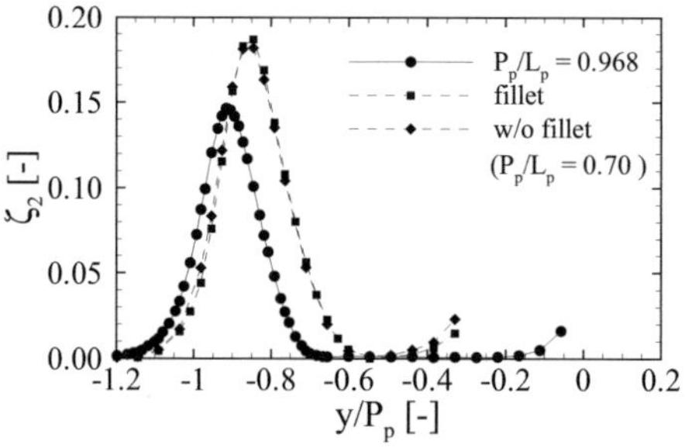

Figure 8.10: Effect of reduced pitch-to-chord length ratio on the trailing edge wake in terms of total pressure loss coefficient for $\beta_{1,ms} = 94.8°$, $s/L_P = 1.03$

The enhancement of the secondary flow's vertical movement, induced by the fillet, is more clearly seen in the plots for the flow-angle deviation. Near the endwall, both curves for large and small pitch ratio with a fillet applied match a maximal overturning of -5.4°. The spanwise extent of the region with highly pronounced under- and overturning spreads slightly more toward midspan for the large pitch ratio. Thus, maximal underturning is also located farther distant from the endwall and is slightly more pronounced. It yields about $\Delta\beta = 2.8°$ for the large pitch-to-chord ratio of 0.968 instead of $\Delta\beta = 2.0°$ as obtained for the small pitch ratio. Removing the fillet, significantly decreased overturning of $\Delta\beta =$ -3.0° can be observed close to the endwall. This indicates sufficiently suppressed cross flow movement from PS to SS. Above a wall distance of 7 %, the evolution of the flow angle deviation is similar for both fillet configurations at the small pitch ratio. The curves reveal similar magnitudes of maximum underturning. It becomes apparent that the fillet influence is limited to the near-wall region only promoting the overturning of the cross flow from PS to SS. With increased loading, the influence of the fillet is continuously reduced (loading results can be found in the appendix A.4.

8.1.5 Airfoil Pressure Distribution

The measurement results of the airfoil pressure distribution at the midspan section of the airfoil are shown in Fig. 8.11 incorporating the effect of increased vane loading. For the reference

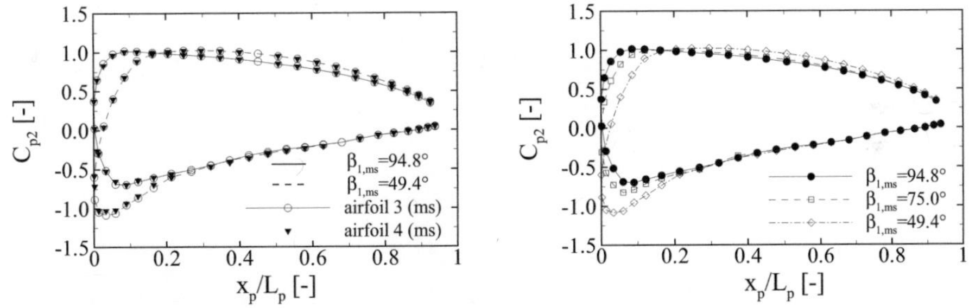

Figure 8.11: Comparison of static pressures for the adjacent vanes 3 & 4 of the center passage and profile pressure distributions in terms of pressure coefficient C_{p2} at airfoil midspan $z/z_{ms} = 1.0$ with variation of loading

case, the distribution reveals loading in the leading-edge part of the profile. The pressure and suction side maximum is obtained at nearly the same position at $x_p/L_p = 0.09$. As the cascade incidence is increased, the loading of the profile is shifted more toward the pressure side which results into stronger acceleration of the flow around the leading edge. Downstream from that position, the flow is decelerated. Therefore, the depth of the suction side maximum is extended to C_{p2} = -0.94 and is determined closer to the leading edge at $x_p/L_p = 0.03$. The pressure side maximum is determined farther downstream of the leading edge at 16 % chord length. The enhancement of the profile pressure gradient between the PS and SS maximum is found to about 21 % with increase of loading.

Furthermore, both airfoil pressure distributions of the vanes 3 and 4 which define the center passage are obtained during the same test run. The static wall pressures are compared in terms of the pressure coefficient C_{p2}, as displayed in the left diagram of Fig. 8.11. Overall good agreement between both pressure distributions is found along the airfoil chord. Differences in pressure reading are estimated between 0.4 and 2.1 % on PS and SS with regard to airfoil 3. The differences significantly increase up to 10 to 15 % when the measured pressure difference approaches zero in the rear part of the pressure side. According to this, both airfoils reveal identical surface pressure distributions, hence similar loading.

8.1.6 Endwall Static Pressure Field

To evaluate the driving forces of the secondary flow structure, measurements of the endwall static pressure field under various loading conditions are conducted. Furthermore, these results are used to evaluate the coolant pressure ratio and the injection parameters as cooling-flow rate, blowing rate and momentum flux ratio per hole with variation of loading. The measurements were conducted on a flat endwall plate without film-cooling holes. The influence of the hole exits on the surrounding pressure field can not be determined. The experimental results for the three test cases of $\beta_{1,ms} = 94.8$, 75.0 and 49.4° are shown in Fig. 8.12 in terms of the dimensionless pressure coefficient C_{p2}. The hole exit locations are plotted for reference as single points where the hole axis emerges through the endwall.

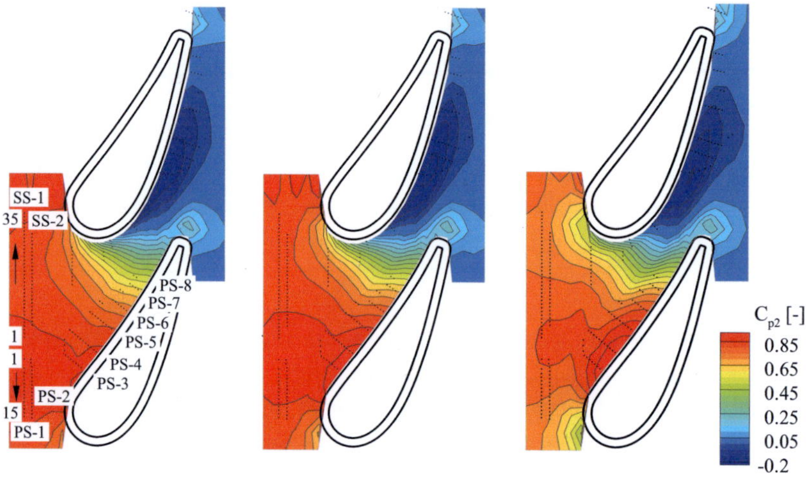

(a) $\beta_{1,ms}$=94.8 deg. (b) $\beta_{1,ms}$=75.0 deg. (c) $\beta_{1,ms}$=49.4 deg.

Figure 8.12: Endwall pressure field in terms of pressure coefficient C_{p2} with variation of inlet flow angle

Observing the endwall pressure field for the vane's operating point $\beta_{1,ms} = 94.8°$, the potential flow effect on the incoming flow is clearly visible upstream of the cascade. It leads to changing static wall pressures across the film rows located at the cascade inlet. Hence, the coolant pressure ratio is significantly altered which influences the share of coolant flow and the local blowing rate for the individual holes in such film rows. The stagnation point is found being placed in the endwall region near the intersection of pressure side and leading edge radius, as indicated by very high values of the pressure coefficient. The flow is accelerated toward the pitchwise pressure gradient across the vane passage.

The maximum pressure difference of 2440 Pa is obtained between the PS at 10% chord length and SS trailing edge at approximately 60% chord length. The lowest pressures of C_{p2} = -0.2, hence the largest mainflow velocities are observed at the suction side downstream of the vane throat. The flow is accelerated along the pitchwise pressure gradient toward the vane throat. Downstream of that point, the flow is decelerated. A region of slightly higher pressures is seen closely downstream of the trailing edge near the airfoil fillet at y/P_P = -0.11 and $x/P_P = 0.03$. It may be attributed to the disturbed flow field induced by the fillet.

As the incidence increases, the stagnation point is shifted along the pressure side toward the trailing edge as seen in Fig. 8.12. At maximum loading, the location of the stagnation point is found at app. 35 % chord (near PS-4, $y/P_P = 0.52$, x/P_P = -0.47). On one hand, it results into stronger acceleration of the flow around the leading edge toward the suction side. This gives lower wall pressures for the holes placed around the leading edge. On the other hand, the movement of the stagnation point changes the direction of the pressure gradient that drives the near-wall crossflow. It redirects more in lateral direction which leads to stronger streamline curvature toward SS and earlier detachment of the cross flow from the pressure side. Due to this, significant change in the wall pressure field is seen at the cascade inlet and in the vane passage. The potential flow effect on the incoming flow is less pronounced with increased loading which leads to reduced wall pressure for the inlet rows SS-1 to PS-2. The wall pressures for rows that are located at the suction side near the leading edge (as eg. SS-3 to SS-6) are significantly reduced due to enhanced flow acceleration. The depth and extent of the pressure minimum near the suction side is found to be slightly increased. The effect of loading on the wall pressure field is less severe for the rear part of the endwall. The region of higher wall pressures downstream of the trailing edge in the flow path of its wake is not altered with loading.

8.1.7 Comparison to Results of the Numerical Computations

Airfoil Pressures. Detailed numerical simulations were performed on the cascade flow field without coolant injection. The computational results are used for the validation purposes to correctly capture the overall main flow and the near-wall flow pattern caused by the secondary flow. Therefore, selected numerical results are compared to the experimental data obtained during the aerodynamic performance investigations. First, numerically computed airfoil pressures at vane midspan are drawn versus the experimental data obtained at the same spanwise section. Both curves are plotted in Fig. 8.13 in terms of the dimensionless pressure coefficient C_{p2}. The data are shown with regard to the dimensionless coordinate x_p/L_p of the profile coordinate system. The comparison is done for the reference case and the test cases with increased loading.

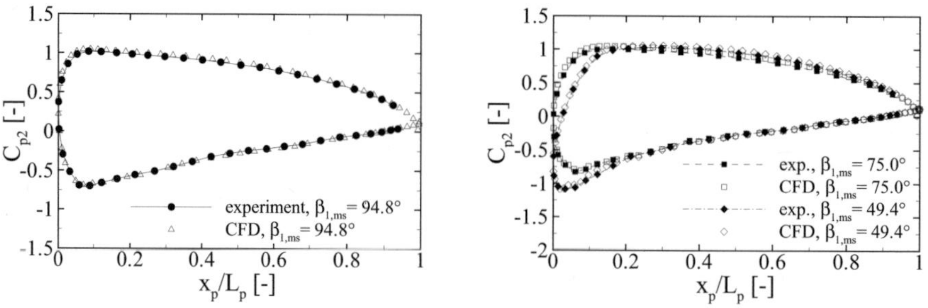

(a) reference test case (b) increased loading conditions

Figure 8.13: Comparison of CFD-data and experimental results for the airfoil pressure distributions at vane midspan with variation of cascade loading

It can be seen clearly that the curves for the airfoil pressure computed by the CFD coincide with the measurement results. Only negligible deviations can be observed that lie within the error margin of the wall pressure's measurement uncertainty. For the reference case, the CFD-computed distribution reveals a similar front-loaded profile with its maximum pressure difference located at similar chordwise positions close to the leading edge. Since the maximum pressure difference from PS to SS attributes to the vane loading and is the driving force for the cascade flow, accurate prediction is essential.

Comparing CFD and experiments for the reference test case, the position and the depth of the SS pressure maximum is captured very well at the same location $x_p/L_p = 0.09$. But also the pressure side maximum is matched in its axial location and depth. Increasing the incidence will shift the stagnation point and hence the loading toward the pressure side. That results into stronger acceleration of the fluid around the leading edge. Due to this, the depth of the SS maximum is extended and moves in upstream direction. It results in an increased maximum pressure difference between PS and SS. This behavior is also predicted very well by the numerical simulations. Hence the numerical and experimental airfoil pressure distributions show excellent qualitative but also quantitative agreement.

Endwall Pressure Distribution. For initiating the near-wall cross flow and the passage vortex, the development of the pitchwise pressure gradient is important. The endwall pressure field should reveal informations about the evolution of the driving forces for the secondary flow with change of loading. Hence, its correct prediction is essential to capture the flow effects. Due to that, the validation of the numerical results is extended on the endwall pressure distribution. The comparison is shown in Fig. 8.14 by means of the isentropic Mach number.

Considering the reference case with an inlet flow angle of $\beta_{1,ms} = 94.8°$, the isentropic

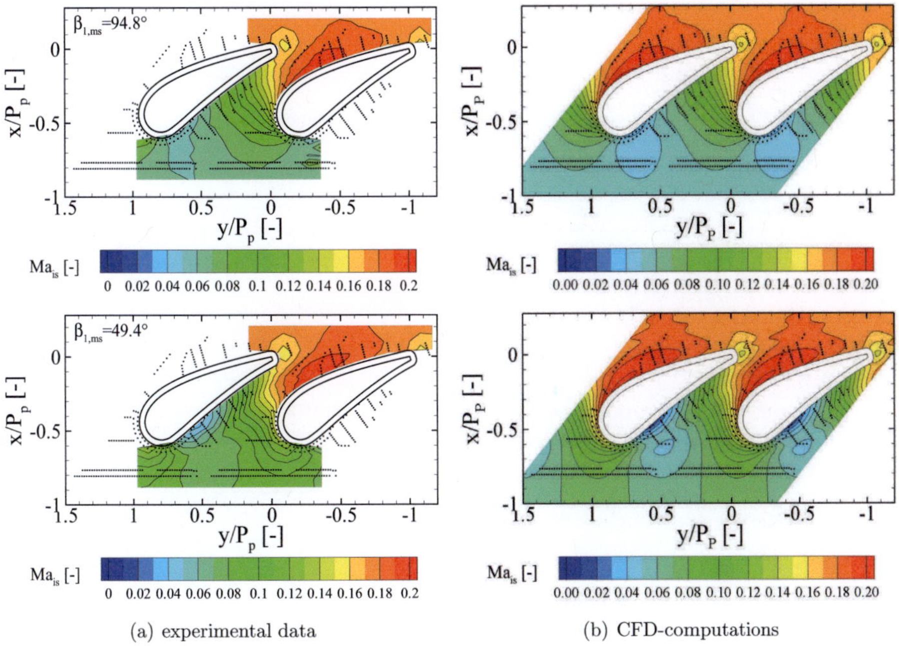

Figure 8.14: Comparison of CFD-data and experimental results for the endwall static pressure distribution in terms of the isentropic Mach Number Ma_{is} with variation of loading

Mach number reveals quite good qualitative agreement between the CFD and experiments. The location of the Mach number's minimum (equals the pressure maximum) that represents the stagnation point is matched quite well in its position and depth close to the leading edge. The spatial extent in upstream direction is slightly underpredicted in the numerical computations. The Mach number maximum (pressure minimum) at the suction side is found to agree reasonably well with the CFD-results with respect to its location close to SS in the rear part of the passage.

The depth of the maximum peak close to the SS-wall and its spatial extent is slightly overpredicted by the CFD-model. Remarkably, the region of locally increased isentropic Mach number downstream of the fillet is captured reasonably good in its spatial extent and its peak value. At elevated loading, similar trends are observed for the Mach number maximum (pressure minimum). In contrast to the good agreement of the prediction for the stagnation region at the reference case, the computational results show a significantly enhanced stagnation region computed by the numerical simulations. Although the location of the pressure maximum

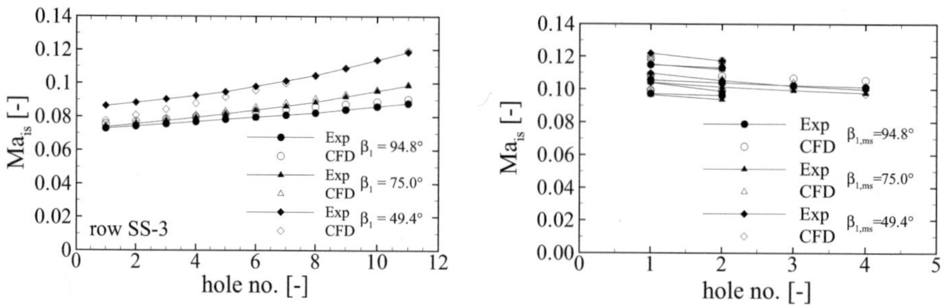

(a) row SS-3 (b) row PS7-PS8

Figure 8.15: Comparison of CFD-data and experimental results for the endwall static pressure at the injection location of two different film rows SS-3 and PS7-PS8 by means of the isentropic Mach number

agrees quite well, numerical simulations show an overpredicted spatial extent in pitchwise and upstream direction. Due to that, deviations in the wall pressure are obtained in the front part of the passage between the CFD-model and the experiments.

To discuss this issue in detail, isentropic Mach numbers are extracted at some hole exit locations. The results are displayed in Fig. 8.15 for the film row SS-3 and the combined row PS7-PS8. Considering the reference case, overall good agreement between the experiments and CFD is observed across both film rows. At maximum loading, the isentropic Mach numbers still show quite good agreement for holes of row SS-3 that are positioned close to SS (hole numbers 6-11). Thus, it indicates that the acceleration around SS is matched well by the numerical simulation. Instead, increased differences are observed for the outer holes (1-5) of row SS-3. Here, the overprediction of the stagnation region takes effect by increasing the endwall pressure which then leads to lower Mach numbers. Similar behavior can also be observed for other rows (e.g. SS-1/2 or PS-1/2) that are located in regions where the wall pressure is affected by the stagnation region. Further downstream of the stagnation point toward the rear part of the passage, the deviations between CFD and experiments become visibly smaller. Hence, the numerical results at the hole exit locations of rows PS7-PS8 are in very good agreement with the experiments even for high airfoil loading.

Total Pressure Losses. The total pressure losses are compared between the experiment and the numerical simulations in terms of the spatial distribution of the loss coefficient ζ_2 at a streamwise distance of $s/L_P = 1.03$ and the area-averaged loss coefficient $\overline{\overline{\zeta}}_2$. Comparing the local total pressure losses, as displayed in Fig. 8.16, it becomes clear that the numerical simulations quantitatively over-predicted the loss generation induced by the secondary flow structures. Here, the distinctive loss peaks related to the passage and the trailing edge vortex as well as the loss maximum close to the endwall are considerably overrated. The used numerical model lacks from overpredicting the losses that are due to turbulent mixing and dissipation. Due to that, the production of turbulence is overrated by the used eddy-viscosity turbulence model primarily in the endwall-suction side corner. At this location, the near-wall cross flow interacts with the newly-grown profile boundary layer which greatly attributes to the loss generation. Application of a Reynolds-stress model solving transport equations for each Reynolds stress term should further improve the prediction of the total pressure losses with the used CFD-model.

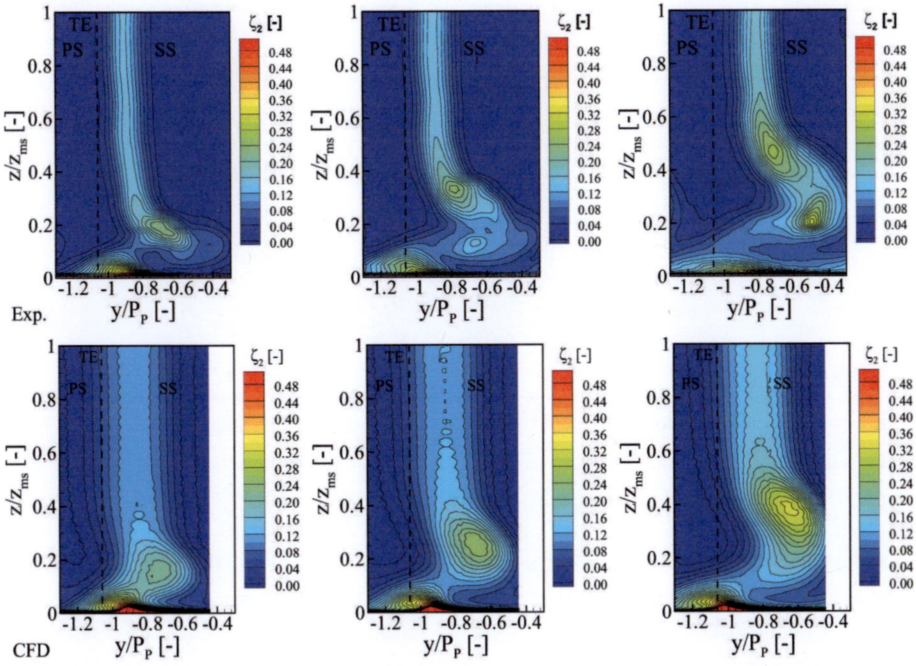

(a) $\beta_{1,ms} = 94.8°$ (b) $\beta_{1,ms} = 75.0°$ (c) $\beta_{1,ms} = 49.4°$

Figure 8.16: Comparison of experimental results (upper row) and CFD-data (lower row) for the total pressure loss coefficient ζ_2 at cascade exit (MP4, $s/L_P = 1.03$) with variation of loading

Nonetheless, the typical "dog-bone" shape of the endwall-loss region, the position of the loss peaks and the pitchwise position of the midspan wake was matched in the simulation. Also, the qualitative trend of the loss region's growth in lateral and spanwise direction and the increase in wake depth with higher loading is reproduced similarly. In contrast to that, the wake's pitchwise extent is significantly overestimated and leads to a broader trailing edge wake and higher profile losses obtained by the CFD. The treatment of the simulated boundary layers as fully-turbulent contributes to this effect. This leads to greater friction losses for a turbulent boundary layer compared to a boundary layer that undergoes transition due to increased turbulent dissipation.

At real flow conditions, the profile boundary layer starting to grow at the leading edge is often in laminar state and transition from laminar to turbulent sets in when the flow is accelerated in streamwise direction. Furthermore, the increase of the boundary layer thickness along the profile is more pronounced for a fully-turbulent boundary layer compared to a realistic transition from laminar to a turbulent state. This leads to increased losses and laterally-broadened midspan wake. At the time when the simulations were performed no proper transition model was available in ANSYS CFX^{TM} but it might improve the simulation of the profile losses as long as it is able to correctly predict the starting point of the transition and its length along the profile.

inlet flow angle $\beta_{1,ms}$ [°]	experiment $\overline{\overline{\zeta}}_{2,exp}$ [−]	CFD $\overline{\overline{\zeta}}_{2,CFD}$ [−]	deviation $\frac{\overline{\overline{\zeta}}_{2,CFD}-\overline{\overline{\zeta}}_{2,exp}}{\overline{\overline{\zeta}}_{2,exp}}$ [%]
94.8	0.0437	0.0622	+42.3
75.0	0.0520	0.0673	+29.4
49.4	0.0709	0.0831	+17.2

Table 8.2: Comparison between experiments and CFD for the area-averaged total pressure loss coefficients $\overline{\overline{\zeta}}_2$ with variation of loading

Similar trends as for the spatial pressure losses are also observed from the area-averaged quantities as depicted from Table 8.2. In the reference case, distinct differences of 1.85 % points in loss coefficient are seen between the CFD-simulations and experimental data. This equals a difference for the CFD-results of + 42% with regard to the experiment. When vane loading is increased, the deviations become smaller yielding only 1.22 % which equals about +17 % difference between CFD and experiment. Despite the overrating of the total pressure losses about 17 to 42% for each incidence, the trend of loss increase as function of loading is qualitatively reproduced.

Surface Streamlines. For analyzing the film effectiveness distributions in streamwise direction, information about the flow direction near the endwall is required. The interaction of the film jets with the highly distorted main flow boundary layer at their injection locations is highly complicated and of three-dimensional nature. A schematic of the situation for the film-cooling injection on a vane platform is shown in Fig. 8.17. The behavior of the cooling jet normal to the wall, including its bending and attachment to the surface, is subject to several parameters. Although the vector of the coolant momentum, the film hole geometry and other parameters attribute to the injection situation, the main flow momentum vector has tremendous effect on the curvature of the jet trajectories and the interaction of the main flow with the injected coolant in wall normal direction.

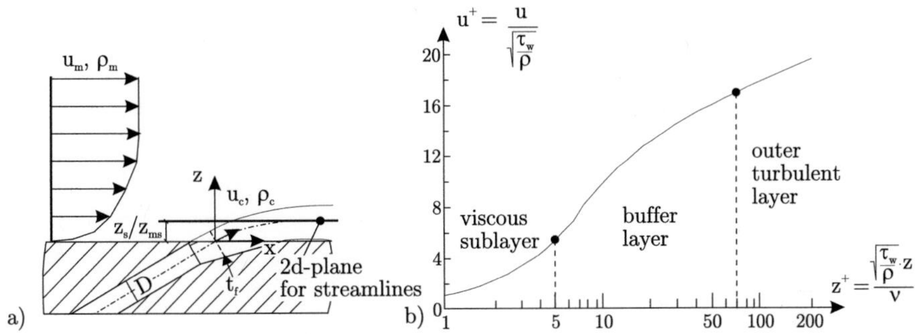

Figure 8.17: Schematic of film injection situation on the endwall (a) and subdivision of near-wall region of a turbulent boundary layer profile (adapted from Schlichting [114]) by means of the dimensionless velocity u^+ and spanwise coordinate z^+ (b)

Since the penetration of the coolant into the crossflow and the bending of the jets is dependent on the coolant and main flow momentum, the jet trajectories are dictated by the injection angle. At low jet momentum and strong near wall cross flow, the jet trajectories and the injection direction would be primarily dominated by the near-wall flow. With increasing coolant momentum, the influence of the cross flow on the injection direction is reduced to let the trajectories being dictated more by the hole axis. Hence, finding streamlines that represent the near-wall flow direction and the streamwise propagation of the coolant is not only affected by the strength of the secondary flow but also by the coolant momentum and the three-dimensional character of the injection situation. Unfortunately, the TSP-measurements deliver only a two-dimensional plot displaying the temperature footprints of the film-cooling jets. This two-dimensional temperature distribution is a projected expression of the three-dimensional process of momentum and heat transfer that takes place inside the core of the film jets and within the shear layers at their edges.

Therefore, a two-dimensional plane is more appropriate to analyze the streamwise development of the cooling films. Two different types of streamlines are available in ANSYS CFX^{TM} to visualize the flow direction in space or on a surface of interest. On one hand pathlines can be computed that are defined as a three-dimensional line which a fluid element follows as it travels through the computational space. On the other hand also pseudo-streamlines can be determined on pre-defined flat or curved surfaces that represent the projected velocity components on this surface. To examine the nature of the streamwise film behavior and the endwall film effectiveness under the influence of the secondary flow, two-dimensional surface streamlines are determined on spanwise surface planes that are oriented parallel to the endwall.

Extensive post-processing was applied to the CFD-simulations for each loading configuration. Two surface streamlines are defined for each hole whereas the one streamline is initiated at the hole discharge and the other half-way between two holes. The streamline coordinates and the computed quantities of the flow field were extracted for all film rows and each loading case. The individual streamlines are then combined for each film row to be superimposed onto the measured temperature distribution. To select an appropriate wall distance for the analyzing plane, the computed streamline pattern is compared to surface flow visualizations using a mixture of oil and carbon-black powder and to the TSP-temperature distribution for a range of wall distances. Furthermore, some restrictions must be taken into account not to place the plane inside a region of the main flow boundary layer that is highly influenced by viscous effects.

In order to physically describe the shape of the velocity profile inside the boundary layer, the near-wall region can be subdivided into three distinct regions which are characterized by its dimensionless wall distance. This subdivision of the near-wall region is illustrated in Fig. 8.17 plotted in semi-log coordinates. Within the innermost layer, also denoted as viscous sublayer, the viscous shear stresses, that are independent from the wall distance, dominate an almost laminar flow regime. In this wall region, molecular viscosity plays a major role in momentum and heat transfer. The mean velocity is dependent on the fluid properties, wall shear stress and distance. In the outer layer, denoted as fully-turbulent region, the shear stress monotonically decreases toward zero. Here, turbulence plays a dominate role over viscous effects. Finally, an interim region, also denoted as buffer layer or blending region, can be identified between the viscous sublayer and the outer turbulent region where effects induced by molecular viscosity and turbulence are of equivalent importance. Taking this into account, it is required to place the analyzing plane outside of the viscous sublayer and the blending region to avoid viscous effects on the near-wall streamlines. Within the fully-turbulent layer, the streamline trajectories are independent of the wall distance as shown in Fig. 8.18.

The surface streamlines are then extracted from a surface cut-plane placed parallel to the endwall. The wall distance of this plane was varied starting at the first node of the computational grid at $z/z_{ms} = 0.00007$ until a normalized wall distance of $z/z_{ms} = 0.01$ far outside of the viscous sublayer and the blending region. The surface streamlines are then superimposed onto the flow visualization pattern of the near-endwall flow obtained from the oil-carbon film technique. Since the wall stress is connected to the velocity gradient normal to the wall, the oil-film pattern may indicate the flow direction in the near-wall region. Due to the large velocity differences between the mainflow and the applied oil-carbon black mixture, the surface of the mixture that is exposed to the flow can be assumed as solid wall, although it is a slow moving fluid. The mixture is applied as finite layer onto the endwall with an approximate thickness of

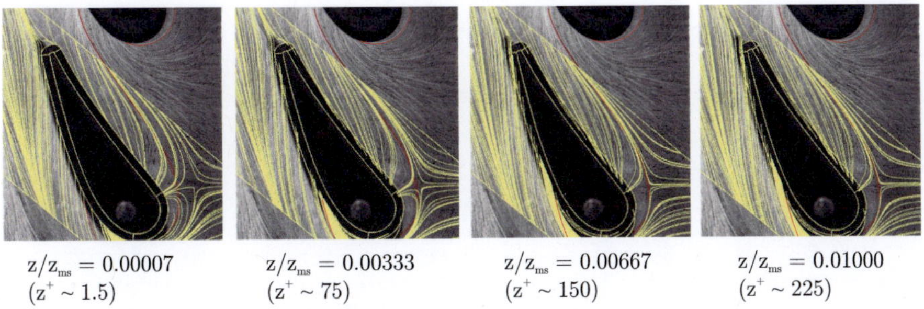

Figure 8.18: Comparison of CFD-computed surface streamlines at varying spanwise distance $0.00007 < z/z_{ms} < 0.01$ superimposed onto the oil-carbon film pattern on the endwall derived from flow visualization studies for the reference test case at $\beta_{1,ms} = 94.8°$

$0.5\,mm$ - $1.0\,mm$ and the wind tunnel is operated for 15 minutes. The resulting oil-film flow pattern is then photographed and shown in Fig. 8.18 along with the superimposed streamlines at different spanwise distances. The data is plotted for the reference case with an inlet flow angle of $\beta_{1,ms} = 94.8°$.

The non-dimensional wall distance for each plane with respect to the wall shear stress τ_w is then determined from an area-averaged z^+ at the first cell node of the CFD-mesh. Thus, it is determined with $z^+ = 1.50$ for the reference case. Assuming the wall shear stress τ_w to be constant across the boundary layer, the range of the plane's non-dimensional wall distance can be obtained within $1.50 < z^+ < 225$. The plane nearest to the endwall is located at $z/z_{ms} = 0.00007$ directly placed within the viscous sublayer as indicated by its non-dimensional wall distance $z^+ \sim 1.5$. Thus, large differences in the flow direction between the oil-film pattern and the CFD-streamlines are clearly visible at the leading edge. The potential flow effect caused by the leading edge extends more in upstream direction, resulting in enhanced deflection of the flow around the leading edge. Significant shift of the saddle point in upstream direction causes earlier diversion of the streamlines. Hence, the streamline pattern at this plane does not match the oil-film flow pattern and can not be used for the intended data analysis purposes. The flow field is rather dominated by viscous effects changing significantly the streamline curvature.

Increasing the wall distance up to $z/z_{ms} = 0.0033$, the streamline pattern still reveals distinct changes in the near-wall flow direction. The position of the saddle point shifts closer to the leading edge reducing the potential flow effect to the incoming main flow boundary layer. Between this wall distance at 0.3% midspan position and the plane at 0.7 % midspan at $z/z_{ms} = 0.0067$, only minor differences are observed for the streamline pattern. This is due to the plane's position at the lower border of the fully-turbulent region (at $z^+ = 150$) where the turbulence dominates over the viscous effects.

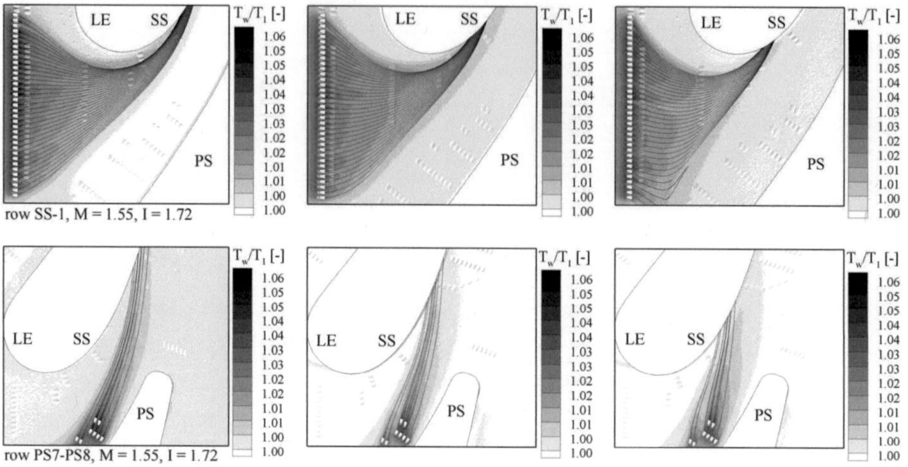

(a) $\beta_{1,ms} = 94.8°$ (b) $\beta_{1,ms} = 75.0°$ (c) $\beta_{1,ms} = 49.4°$

Figure 8.19: Superposition of computed streamlines at a wall distance of $z/z_{ms} = 0.0067$ onto the TSP-acquired temperature distributions with variation of loading, film-cooling injection with row SS-1 (upper row of images) and combined row PS7-PS8 (lower image row)

Further increase of the wall distance to $z/z_{ms} = 0.010$ (equals $z^+ = 225$) would still place the plane inside the outer turbulent layer. Therefore, no recognizable changes in the flow direction are observed. as indicated by the streamline comparison with the oil-film pattern, as displayed in Fig. 8.18. The streamline comparison with the oil-film pattern agrees quite well for the plane at $z/z_{ms} = 0.0067$ and no further changes can be observed toward $z/z_{ms} = 0.01$.

To further validate the correct selection of the plane's wall distance, the 2D-streamlines from the CFD-computations are plotted onto the raw temperature image obtained by the TSP-measurement technique. The results of this analysis are plotted in Fig. 8.19 for the spanwise plane at $z/z_{ms} = 0.0067$. Two film-cooling rows are selected: row SS-1 located at the cascade inlet and PS-7/PS-8 placed near the pressure side in the passage center. Generally, the streamlines show excellent agreement with the temperature footprints of the films for all row cases. Considering increased loading conditions, the near-wall flow field is certainly altered by enhancing the streamline curvature. Also at such flow conditions, the streamwise behavior of the temperature footprints is still reproduced very well by the computed streamline pattern. It is remarkable that such unique features that occur for row SS-1 at maximum loading can be reproduced as well. The influence of the near-wall flow on the coolant's injection and traveling direction due to a shifting stagnation point can be traced by the imposed streamlines. Due to

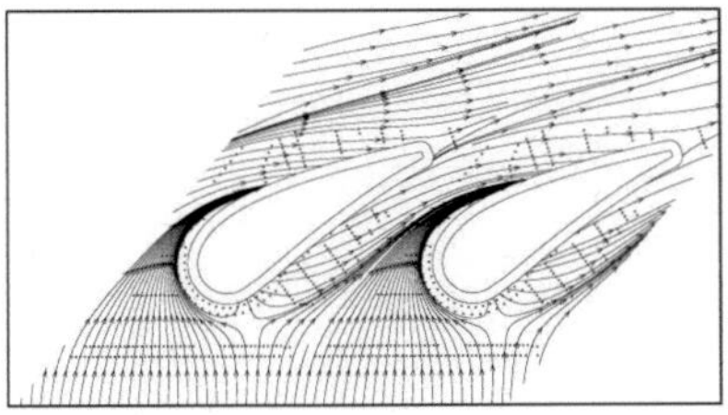

(a) $\beta_{1,ms} = 94.8°$

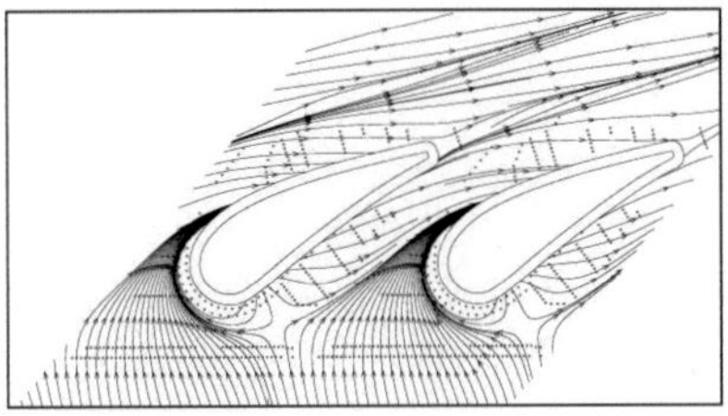

(b) $\beta_{1,ms} = 75.0°$

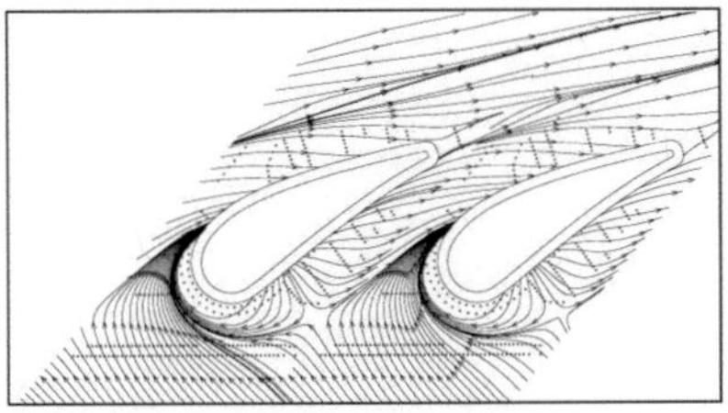

(c) $\beta_{1,ms} = 49.4°$

Figure 8.20: CFD-Streamtraces at a spanwise wall distance of $z/z_{ms} = 0.0067$ with variation of loading

that, the observed dislocation of the saddle point of the incoming flow leads to opposite diversion of film-cooling jets that are injected from the first 5-6 holes toward the pressure instead of the suction side.

Finally, the spanwise distance z/z_{ms} of the streamline plane is specified combining the very good agreement of the streamlines with the oil-film flow visualizations and the coolant trajectories visible in the wall temperature fields. Furthermore, it should be noted that the films follow a three-dimensional trajectory in space and have a finite lateral and spanwise extent which broadens in streamwise direction. Directly at the hole exit, their lateral spread may be equivalent to the maximum width of the diffuser exit. The spanwise extent of the film is dependent on the hole geometry, the blowing parameters and how the films interact and mix-out with the main flow.

As a first approximation, the spanwise extent of the cooling films that are initiated from the shaped holes is estimated with the maximum height of the diffuser cross-section. It is formed by an imaginary plane perpendicular to the hole axis at the point where the hole axis emerges through the surface. Hence, it is determined with $H_{fan}/D \sim 1.7$. The position of the jet center is estimated by half of the maximum height which equals to an approximate non-dimensional wall distance of $z/z_{ms} \sim 0.007$. The analyzing plane for the streamlines should be located near the spanwise center of the film to approximate an averaged three-dimensional coolant motion on a 2D-surface.

Bearing in mind that the streamline pattern is influenced by viscous effects below a wall distance of $z/z_{ms} = 0.0033$ and shows reduced changes with greater wall distances, the coolant development in the temperature images is matched well on a two-dimensional plane at a wall distance of $z/z_{ms} = 0.0067$. The extracted streamline pattern from this spanwise plane is used to evaluate the streamwise film effectiveness and is shown in Fig. 8.20 for all loading cases. Additionally, this wall distance is comparable to the finite thickness of the oil film layer. After detailed description of all data reduction steps applied to the raw TSP-images, the film effectiveness distributions and its streamwise development are discussed during the next sections for different endwall film-cooling rows and with respect to changed cascade loading.

8.2 Film-Cooling at Cascade Inlet: Row SS-1

The coolant row SS-1 is located at x/L_{ax} = - 1.39 which is approximately 20 % chord upstream of the leading edge and contains 35 shaped holes (see also Fig. 8.21). At the hole entry, these holes are designed with a branch-like metering part with its axis normal to the endwall surface. After a length-to-diameter ratio for this branch-part of L_h/D_{br} = 3.6, the coolant enters the inclined metering section with an inclination angle smaller than 30°. The compound angle with respect to the axial direction is quite small, aligning the holes of row SS-1 with the incoming flow at operating conditions. At the hole intersection between branch-part and inclined metering section, the coolant experiences a deflection in flow direction of about 70°. As the coolant is injected into the approaching cascade boundary layer, the diffuser provides sufficiently enhanced lateral spreading due to its moderate lateral and axial opening angle.

8.2.1 Coolant Injection

With the endwall static pressure field showing spatial gradients due to the acceleration of the main flow, it also varies with airfoil loading. To maintain consistent coolant injection with changing incidence, the observed effects on the static wall pressure must be considered in the calculation of the blowing parameters. Therefore, the endwall pressures for each loading test case are determined at the location of the hole breakouts where the hole axis emerges through the endwall. A comparison of the pressure coefficient C_{p2} for varying incidence at the injection locations of row SS-1 can be depicted in Fig. 8.21.

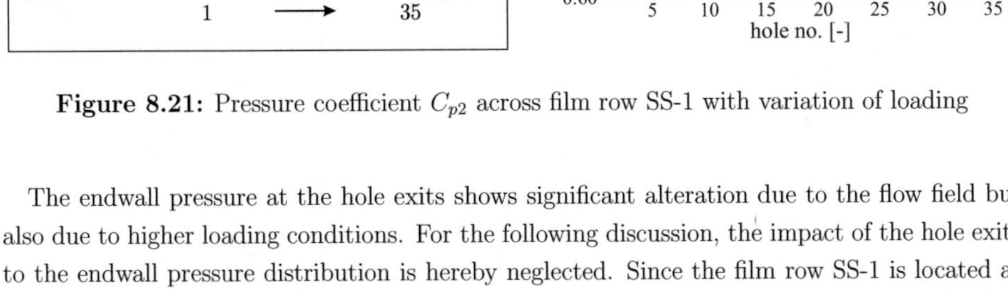

Figure 8.21: Pressure coefficient C_{p2} across film row SS-1 with variation of loading

The endwall pressure at the hole exits shows significant alteration due to the flow field but also due to higher loading conditions. For the following discussion, the impact of the hole exits to the endwall pressure distribution is hereby neglected. Since the film row SS-1 is located at the cascade inlet, the potential flow effect caused by the large leading edge is clearly visible in the static pressure distribution. The stagnation region located at approx. 9% chord length (operating point with $\beta_{1,ms}$ = 94.8°) causes higher pressures and therefore an increase in the pressure coefficient C_{p2} for the first third of the film holes. The maximal value is obtained at hole 1 with $C_{p2} = 0.867$ and gradually decreases with greater hole number. Hence, the hole positions get closer to the suction side of the next vane and the acceleration of the main fluid around the leading edge causes lower static wall pressures. The minimum of the pressure coefficient $C_{p2} = 0.838$ can be found between hole 18 and 22 where it increases again toward hole 35 as the next airfoil with its stagnation point is approached.

With moderate increase of the incidence to $i = 15°$, the differences in endwall pressure is less severe for the first holes. This results into equivalent values for the pressure coefficient compared to the operating point and is mainly attributed to the less significant shift of the stagnation point along the pressure side, as visible in Fig. 8.12. Slightly increased acceleration around the leading edge of the next vane causes stronger decrease of the wall pressure with greater hole number. The region with lowest C_{p2} of about 0.822 is shifted toward hole 21 to 24 and is less broaden than for the reference test case with an inlet flow angle of $\beta_{1,ms} = 94.8°$. These effects are further enhanced with increasing the incidence to $i = +40°$. The results show a significant drop in the pressure coefficient level due to the increased inlet main flow velocity and the stronger acceleration of the flow around the airfoil. The maximum of C_{p2} is found still to be close to the first three holes with lower magnitudes of 0.811. The plateau of minimal endwall pressures with $C_{p2} = 0.73$ is obtained to be moved further toward the end of the film row and between holes 27 to 32, respectively.

The endwall pressures have significant influence on the coolant-to-main flow pressure ratio and therefore on the coolant mass flow. Hence, the individual mass flow rates per hole are calculated with the given equations in chapter 5.2. The resulting mass flow distribution and the corresponding hole blowing rates for row SS-1 are shown in Fig. 8.22.

Figure 8.22: Distribution of coolant mass flow rate $\dot{m}_{c,i}/\dot{m}_c$ (left) and normalized hole blowing rate M_i/M_{av} across row SS-1 with variation of loading

For the reference case at minimum loading, the highest mass flow rates occur at the hole locations with the lowest pressures. This is clearly attributed to a rise of the coolant pressure ratio across a specific film hole when the static wall pressure decreases. The data for $\beta_{1,ms}$ $= 94.8°$ reveals a nearly symmetric evolution with regard to the holes 18 to 22 with varying hole mass flow ratios between 2.8 and 2.9 % of the total mass flow rate supplied to the row. With increased loading, the region of highest hole mass flow rates is shifted toward greater hole numbers and therefore toward the next airfoil's suction side. This leads to an asymmetric distribution of the coolant mass flow which is attributed to the downstream shift of the stagnation point along the pressure side. Furthermore, slightly enhanced differences in hole mass flow rates within 2.7 to 2.94 % are observed for the loading case of $\beta_{1,ms} = 75.0°$. The deviations decrease again at maximum incidence ranging from 2.8 to 2.9 % of the total mass flow rate. This is due to the slightly increased inlet flow velocities which partially compensate the influence of the potential flow effect.

Fig. 8.22 reports the distribution of the hole blowing rate with variation of loading. A asymmetric evolution around hole 21 is observed for all loading conditions with a constant hole blowing rate about 4 % lower than the averaged row blowing rate. Thus, the lowest blowing rates occur at the hole locations with minimum endwall pressures and vice versa. The change in endwall pressure across the row results only in a moderate change in coolant mass flow rate of about ± 0.1 %. The effect on the hole blowing rate is therefore mainly attributed to the changing mass velocity of the main flow across the film row. Decreasing endwall pressure leads to an increased coolant pressure ratio and slightly increased mass flow rates across the hole which is balanced out by the increasing mass velocities of the main flow. For the reference test case at minimum loading, the local blowing rate for the holes 1-20 increases about 11 % with decreasing hole number. The maximum blowing rate is seen for hole 1 with $M_i/M_{av} = 1.07$. The smaller part of the row with the holes 22-35 experience also an increase in blowing rate of about 10 % leading to $M_i/M_{av} = 1.06$. The deviations of the local hole blowing rate with regard to hole 21 become more pronounced with higher incidence. The holes 1-20 show enlarged differences up to 15 % at maximum cascade loading, the local blowing rate drops about 3 % for the holes 22-35 located closer to the next airfoil. Due to the enhanced fluid acceleration around its leading edge with increased loading, the mass velocity of the main flow is increased resulting into lower hole blowing rates.

8.2.2 Adiabatic Film-Cooling Effectiveness

Endwall film-cooling experiments on row SS-1 were conducted with a row blowing ratio of $M = 1.55$ and a coolant-to-main flow density ratio $D_R = 1.40$. The coolant is heated to a temperature ratio of $T_R = 1.088$. These film-cooling parameters were maintained for all three loading cases which then result into a momentum flux ratio of I = 1.716 and with regard to the exit area ratio of $I/A_R^2 = 0.092$. The FE-approach to correct for the heat conduction effects is applied to the measurement results as described in chapter 6. The results of the normalized adiabatic film effectiveness η_{aw}/η_{ref} obtained for injection through SS-1 is shown on the left-hand side of Fig. 8.23 with variation of loading. The adiabatic film effectiveness is referred to a reference value η_{ref} which describes the film coverage on the endwall directly downstream of the hole exits. It is defined as the ratio of the maximum width of the diffuser exit divided by the hole pitch. The distributions are superimposed by the computed 2D-streamlines visualizing the simulated near-wall flow direction. Furthermore, streamwise film effectiveness is plotted versus the normalized streamwise distance s/D on the right-hand side of the Fig. 8.23.

In general, the highest normalized film effectiveness can be obtained close to the film injection locations in all incidence cases. The normalized film effectiveness steadily decreases in streamwise direction but is still showing elevated values near the injection location of row SS-3. The films develop into a complete and closed coolant layer and is transported in downstream direction toward to airfoil suction side. As a result of the enhanced lateral spreading caused by the diffuser and the moderate pitch-to-diameter ratio, the film layer develops without any noticeable gaps and the coolant is able to cover about 67 % of the hole pitch. The influence of the near-wall flow field is clearly visible by distorted foot prints of the film-cooling jets which is a result of the deflection of the jets away from their initial injection direction starting very soon at the hole exits. The jets remain attached to the surface over the entire traveled streamwise distance indicating no lift-off from the surface. This is mainly attributed to the diffuser effect reducing the coolant velocity inside the hole and therefore the jet momentum at the hole exit.

Observing the results for the reference case at an inlet flow angle of $\beta_{1,ms} = 94.8°$, the large number of holes that are contained in row SS-1 and subsequently the large amount of coolant mass flow provides a near slot-like coolant distribution that covers an endwall area over 55 % of the airfoil pitch. The highest normalized film effectiveness η_{aw}/η_{ref} is obtained to 0.35 close to the film injection locations approximately 15-20 hole diameter downstream. It degrades in streamwise direction at slow rates still showing magnitudes of η_{aw}/η_{ref} slightly above 0.10 at maximum streamwise distances of $s/D = 220$.

The holes of row SS-1 inject directly into the approaching boundary layer that is rolled up further downstream at the leading edge to form the horse-shoe vortex. It is composed of two different parts, denoted as suction side and pressure side branch, which are transported into the adjacent passages. The PS-branch detaches from the airfoil pressure side and is entrained into the developing passage vortex. However, the SS-branch stays close to the airfoil suction side and proceeds downstream. The rotational movement of the vortex back in upstream direction and its closure due to continuity reasons results into lift-off of the vortex from the endwall surface forming separation lines. The coolant, that is injected upstream of the endwall region around the leading edge, is additionally influenced by this horse-shoe vortex. The flow structure of the vortex branches acts as barrier for the injected films along their lift-off lines, preventing them from crossing this border. Taking into account that the region between the lift-off lines of the horseshoe vortex and the vane is characterized by high heat transfer, the cooling with the jets directly aimed toward the leading edge seems to be ineffective. It leaves this endwall region virtually unprotected which is indicated by very low film effectiveness values. The film effectiveness distribution in Fig. 8.23 shows that the films experience intense diversion

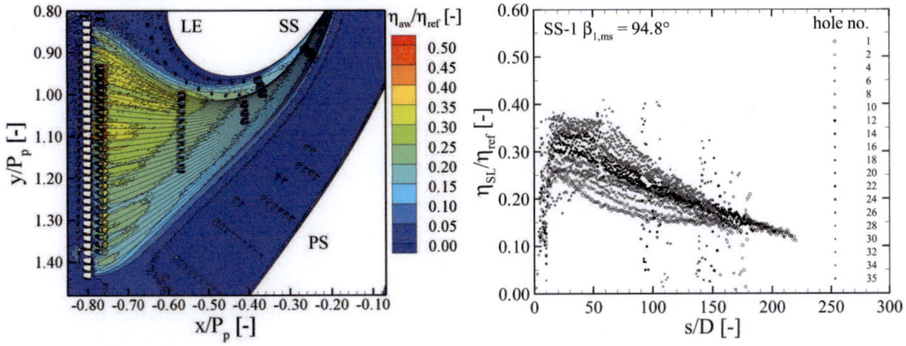

(a) $\beta_{1,ms} = 94.8°$

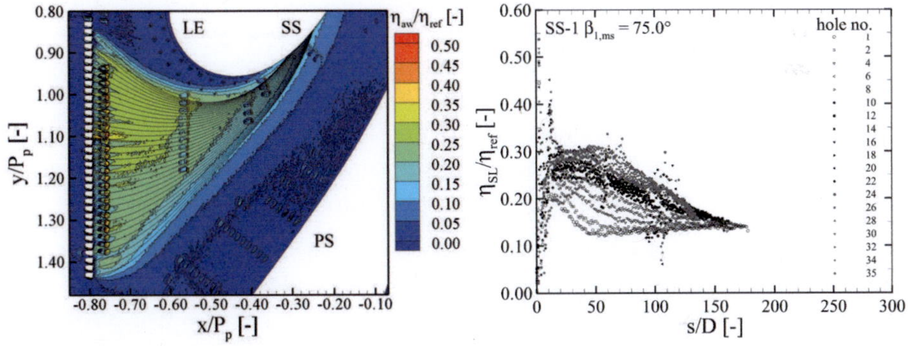

(b) $\beta_{1,ms} = 75.0°$

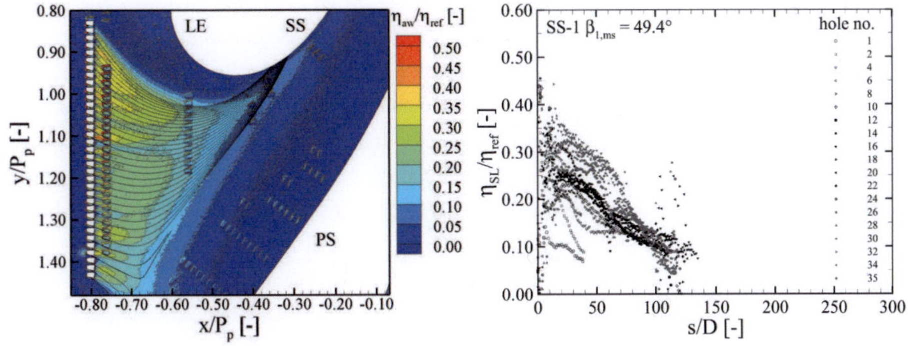

(c) $\beta_{1,ms} = 49.4°$

Figure 8.23: Normalized adiabatic film effectiveness η_{aw}/η_{ref} (left) and streamwise film effectiveness along CFD-calculated surface streamlines η_{SL}/η_{ref} (right) for row SS-1 with variation of loading (a-c) away from their initial injection direction toward the passage center which is prescribed by the orientation of the hole axis. Due to the strong near-wall cross flow that is present at the cascade inlet, the film layer is diverted back to the SS further downstream. It approaches the fillet edge and impinges onto the airfoil suction side at approximately 25 % chord. This position is closely located to the injection location of row SS-6. The coolant is then up-washed from the endwall surface and entrained into the cross flow loosing therefore its cooling capacity.

As the loading is increased to an inlet flow angle of $\beta_{1,ms} = 75.0°$, the diverting effects on the injected film layer become more pronounced. This is mainly attributed to the enhancement of the near-wall cross flow by increasing the pitchwise pressure gradients with larger incidence. The film effectiveness results show that the near-hole flow field and the released film-cooling jets are even more dictated by the cross flow than in the reference case. The location where the film blanket impinges onto the vane suction side is shifted upstream toward the leading edge to approx. 15 % chord length. Due to this, the endwall area that is covered with coolant is slightly reduced. The endwall region around the leading edge, characterized by low film effectiveness, extends in upstream direction which indicates intensified vortical movement of the horseshoe vortex. Additionally, the lift-off line acts not only as a coolant barrier preventing its distribution toward the leading edge but also causes enhanced diversion of the film-cooling jets in pitchwise direction with higher loading. The film effectiveness degrades in streamwise direction at slightly greater rates down to 0.1 leading to reduced maximum streamwise distances of $s/D = 170$.

At maximum loading ($\beta_{1,ms} = 49.4°$), the effects are even more pronounced as for the other incidence cases. The cooling films impinge even closer to the leading edge at 10 % chord length (close to injection location of row SS-4) further reducing the endwall area that is covered by the coolant. Besides, the coolant is forced to distribute along the lift-off line that is created by the pressure side branch of the horseshoe and the passage vortex. The film effectiveness decays at greater rates resulting into values fairly below 0.1 at reduced streamwise distances of $s/D = 120$ where the film layer is then blown off the surface. Remarkably, the films that are released from the first four holes of row SS-1 experience an opposite diversion compared to all other holes. It becomes apparent in the film effectiveness distribution that the footprints of the film-cooling jets are deflected toward the pressure side instead of the suction side. It is a direct result of the displacement of the stagnation point with increased incidence which is shifted downstream along the airfoil's pressure side. Thus, the location of the saddle point changes which results into stronger influence of the dividing near-wall streamlines around the saddle point on the cooling films. At maximum loading, the stagnation point is therefore located between the injection region of row PS-3 and PS-4. The streamwise film effectiveness for these four holes shows lower magnitudes and faster decay in downstream direction persisting until a normalized streamwise distance $s/D = 50$.

8.3 Film-Cooling at Airfoil Suction Side: Row SS-3

The cooling row SS-3 is located on the airfoil suction side at less than 2 % chord and quite close to the leading edge. The film-cooling row is placed in an endwall region that is expected to be influenced by the near-wall cross flow and the horseshoe vortex. The vortex's suction side branch is transported downstream positioning itself in the corner between suction side wall and endwall. This leads to increased turbulence and high heat transfer along with low static endwall pressures due to strong flow acceleration around the leading edge. The film row consists of 11 fan-shaped holes with a total length-to-diameter ratio of 16 for each hole. The diffuser is designed with a moderate opening angle in lateral and streamwise direction. The holes are inclined at slightly less than 30° with regard to the streamwise direction. The pitch-to-diameter ratio P_h/D is 4. The compound angle of the holes is quite small with regard to the axial direction. The dimensionless length-to-diameter ratio of the metering section is about $3/4^{th}$ of the total hole length ratio where the diffusion process takes place over the last three hole diameters.

8.3.1 Coolant Injection

Coolant injection through row SS-3 is subject to significant variation of the main flow field due to its proximity to the leading edge. The endwall pressure coefficient at the hole exit is shown

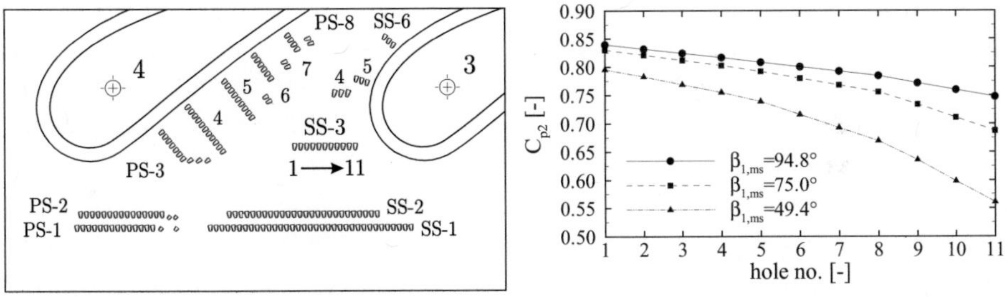

Figure 8.24: Pressure coefficient C_{p2} across film row SS-3 with variation of loading

in Fig. 8.24. It shows significant decrease with larger hole number which is representative of the hole location with regard to the airfoil suction side. The larger the hole number, the closer the hole to the suction side surface. The impact of the hole exits to the endwall pressure distribution is hereby neglected. The effect of the stronger acceleration around leading edge with increasing incidence is clearly visible. The first hole of row SS-3 experiences the highest pressure since it is farthest away from the suction side. The maximum pressure coefficient is obtained at $C_{p2} = 0.84$ and decreases steadily with rising hole number. As the incidence increases, the endwall pressure drops significantly showing greater decrease with increasing hole number. At maximum loading, the pressure coefficient at hole 1 is reduced to $C_{p2} = 0.79$ which furthers decreases across the row about 30 % down to 0.55. This effect is mainly attributed to the shift of stagnation toward the pressure side which results into enhanced acceleration of fluid around the leading edge leading edge and increased main flow velocities at the profile suction side. Therefore, changing the incidence and subsequently the cascade loading has significant impact on the endwall static pressure field as well as on the coolant pressure ratio. Thus, it also influences the share of coolant flow between the holes of row SS-3.

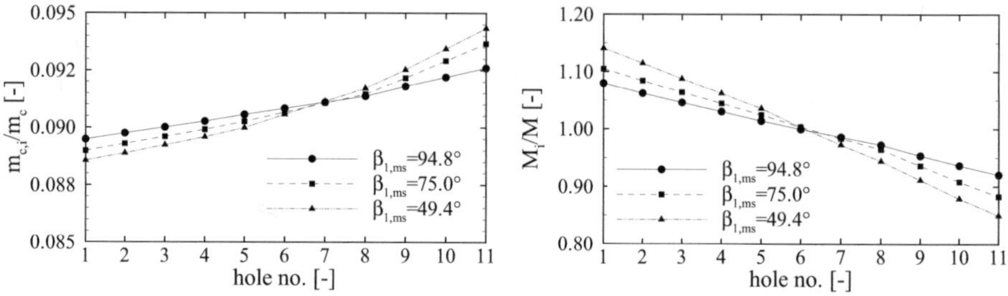

Figure 8.25: Distribution of coolant mass flow rate $m_{c,i}/m_c$ (left) and normalized hole blowing rate M_i/M_{av} across row SS-3 with variation of loading

The comparison of the coolant mass flow rate and the local hole blowing rates for the film row SS-3 is shown in Fig. 8.25. The data is normalized by its averaged value across the row. A slight non-symmetric evolution of the mass flow rate can be observed around hole 7. Slightly increased mass flow rates about 0.15 % are observed for the holes 8-11 closest to the suction side. This is mainly due to the lower static wall pressure increasing the coolant-to-main flow pressure ratio at a constant coolant supply pressure. For the first holes 1-6 the hole mass flow rates are found to be reduced about nearly the same amount yielding differences of 0.2 %. With increasing incidence, the differences of the local hole mass flow rate becomes more pronounced. Hence, deviation of the coolant mass fraction across the row is observed to be slightly increased to ± 0.3 %.

Considering the blowing rate distribution, the evolution has been changed to a nearly symmetric form around hole 6. Higher blowing rates are found for the holes located farther off the airfoil suction side and vice versa. The decreasing endwall static pressures with larger hole numbers lead to increased coolant pressure ratios across the holes. It then results into a slight increase of the coolant mass flow rate per hole but it is balanced out by increased mass velocity of the mainstream due to the strong acceleration around the leading edge. Therefore, the local blowing rate for the holes 6 to 11 decreases about 8 % due to their locations closer to the SS. A similar increase is obtained for the holes 1-6 which are located further away from the SS. The deviation from the averaged row blowing rate becomes more pronounced with higher incidence yielding 14 % difference in the local blowing rate.

8.3.2 Adiabatic Film-Cooling Effectiveness

Endwall film-cooling experiments on row SS-3 were conducted for three loading cases. The developed correction method using the FE-model is applied to the measurement data to obtain the adiabatic film effectiveness. Fig. 8.26 shows the spatial distribution of the normalized adiabatic film effectiveness η_{aw}/η_{ref} with variation of loading. Similar to the results of row SS-1, the adiabatic film effectiveness is referred to a reference value η_{ref} which describes the film coverage on the endwall directly downstream of the hole exits. The row blowing rate is maintained at $M = 1.60$ for all test cases shown. The density ratio for the investigated row is $D_R = 1.40$. The momentum flux ratio is $I = 1.80$. The momentum flux ratio divided by the area ratio between metering and shaped hole section is $I/A_R^2 = 0.096$.

Observing the endwall region downstream of row SS-3, a strong interaction between the coolant and the near-wall flow structure is clearly visible. It can be seen that the film layer is strongly deflected toward the suction side. A complete film layer develops without any noticeable gaps between the jets which is attributed to the enhanced lateral spreading of the coolant caused by the hole diffuser and the moderate pitch-to-diameter ratio. The film jets remain attached to the endwall surface and the film effectiveness results do not indicate any jet lift-off at the investigated blowing rate. The highest film effectiveness is observed closely downstream of the injection location and steadily decays in streamwise direction. As the film layer proceeds downstream, it impinges onto the suction side wall at approximately 25 % chord. The coolant is then up-washed away from the endwall surface being entrained into the vortical movement of the secondary flow.

At higher incidence, these effects are more pronounced due to the intensified secondary flow. This leads to enhanced deflection of the jets away from their initial injection direction. Further, the impingement region on the SS, where the cooling film meets the profile wall, is shifted in upstream direction toward 15 % chord for maximum loading. This means significant reduction of the streamwise distance which directly results into less endwall area covered with films. Therefore, if such off-design flow conditions shall be considered in design practise, the spacing of the film rows must be reduced to avoid uncooled endwall regions. The orientation of the jet trajectories at the hole exits are dictated by the near-wall cross flow which becomes even more pronounced at higher incidence. The CFD-predicted surface streamlines are shown as dashed lines in Figure 8.26 to visualize the near-wall flow direction. The comparison reveals that the behavior of the film layer on the endwall for each incidence is captured quite well by the predicted streamlines.

The streamlines are further used to evaluate the streamwise film effectiveness. The results are plotted versus the normalized streamwise coordinate s/D, shown in the left diagrams of Fig. 8.26 for each loading case. The highest normalized film effectiveness is obtained close to the hole exit for the operating point $\beta_{1,ms} = 94.8°$ which shows remarkable variation of η_{SL}/η_{ref} between 0.3 up to almost 0.5. This lateral gradient is due to the variation in coolant mass

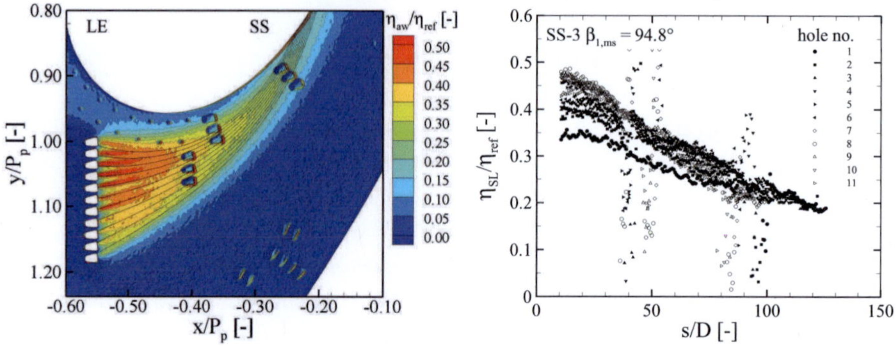

(a) $\beta_{1,ms} = 94.8°$

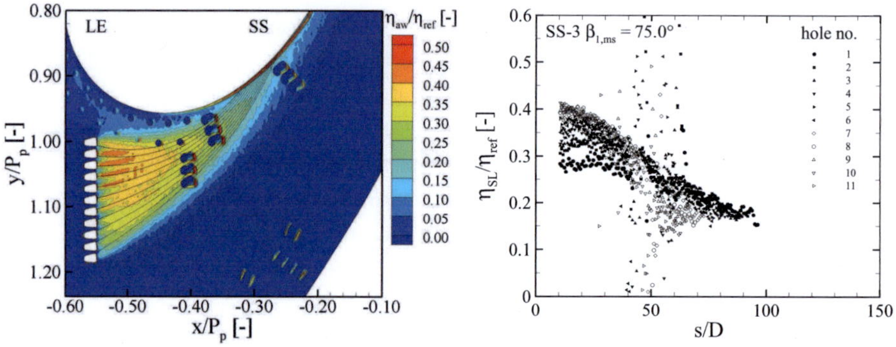

(b) $\beta_{1,ms} = 75.0°$

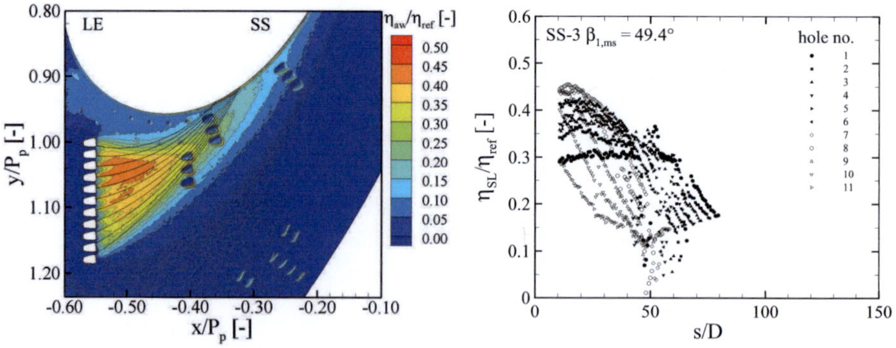

(c) $\beta_{1,ms} = 49.4°$

Figure 8.26: Normalized adiabatic film effectiveness η_{aw}/η_{ref} (left) and streamwise film effectiveness along CFD-calculated surface streamlines η_{SL}/η_{ref} (right) for row SS-3 with variation of loading (a-c) flow across the row SS-3. The inner holes with large hole numbers (6-11) show elevated film effectiveness as a result of the high coolant mass flow that is discharged through those holes and subsequently lower blowing rates. The streamlines of the inner holes (6-11) show enhanced curvature which results into slightly enhanced film effectiveness decay and shorter streamwise distances with s/D = 90. The film-cooling jets that are released from the outer holes (1-5) have to travel larger streamwise distances due to less curved streamlines. The film effectiveness shows less streamwise decay down to $\eta_{SL}/\eta_{ref} = 0.2$ at maximum streamwise distances of s/D = 120 where the film layer then impinges onto the profile wall and is entrained into the cross flow.

With larger incidence and consequently higher loading, the near-wall cross flow that transports the film-cooling layer is intensified. Thus, the film effectiveness at the injection point is slightly reduced and shows steeper gradients in streamwise direction. This indicates intensified mixing and interaction of the coolant with the main flow. The impinging region of the film layer is shifted upstream toward the leading edge due to the more pronounced curvature of the near-wall streamline. At an inlet flow angle of $\beta_{1,ms} = 75.0°$ with moderately increased loading, the films that are released from the outer holes (1-5) show reduced maximum streamwise distances of $s/D = 90$. For the inner holes (6-11), higher streamwise film effectiveness is observed downstream of the hole exit which falls off more quickly until $s/D = 70$. At maximum loading with an inlet flow angle of $\beta_{1,ms} = 49.4°$, these effects are even more pronounced and the intensified near-wall flow leads to enhanced diversion of the film-cooling jets. Although the films injected through the outer holes (1-5) are preserved over larger streamwise distances of 70 to 80 hole diameters and show a more shallow decline of the film effectiveness, the inner holes (6-11) show contrary behavior due to intense interaction of the injected coolant with the near-wall flow. This causes significantly increased film effectiveness decay over shorter streamwise distances of only 50 hole diameters.

8.4 Film-Cooling at Airfoil Pressure Side: Row PS-7/PS-8

The endwall region close to the pressure side is often a hot spot for thermal failure in the gas turbine environment. Here, the near-wall cross flow migrates from the pressure toward the vane suction side to become the passage vortex. The coolant which is injected in this highly three-dimensional and viscous near-wall cross flow experience significant diversion away from the endwall area that is intended to be cooled. This effect was also reported by Friedrichs et al. [38]. Therefore, sufficient platform cooling must be provided along the entire streamwise extent of the airfoil PS. This is often done by placing a lot of holes in this endwall area which disadvantageously increases the total coolant consumption. These influences are expected to be enhanced when the airfoil loading is increased. Out of these reasons, the film-cooling behavior close to the vane pressure side is investigated. Experiments are conducted for the combined holes PS-7/PS-8 to study single row injection. The holes are located at 70 % chord close to the pressure side and 55 hole diameters upstream of the vane throat.

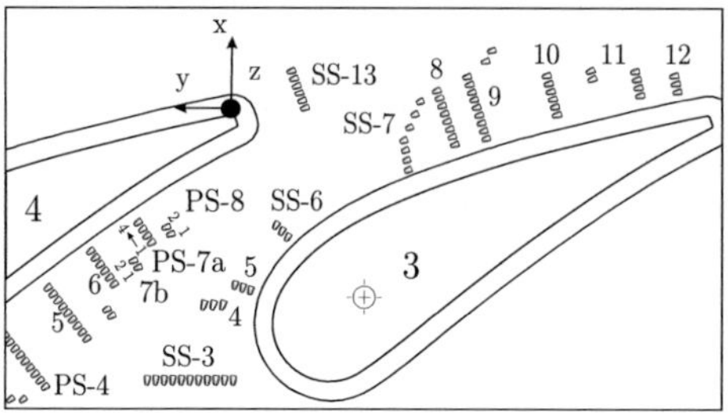

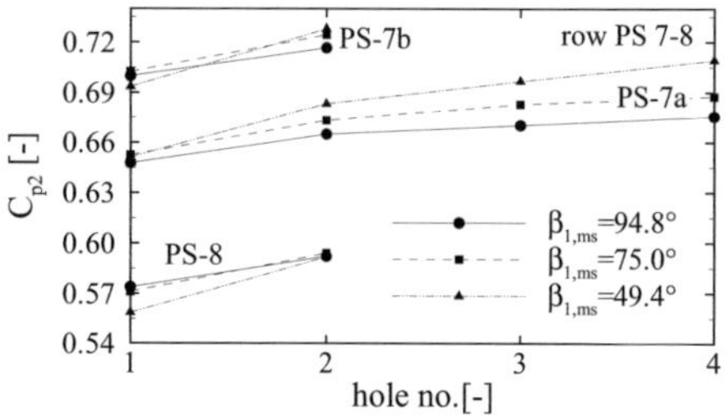

Figure 8.27: Pressure coefficient C_{p2} across film row PS-7/PS-8 with variation of loading

The combined film row PS-7/PS-8 is divided in PS-7a with 4 fan-shaped holes, PS-7b and PS-8 where the latter consists of 2 shaped holes. The second part of row PS-7 is slightly staggered about two hole pitches with regard to PS-7a and is positioned 12 hole diameter upstream of PS-7a. PS-8 is located nearly 10 hole diameter downstream of row PS-7a. Its holes are aligned in pitchwise direction with the first two holes of row PS-7a. The total length-to-diameter ratio of all holes in the combined row is about 16. The diffuser design is similar to the hole geometry of row SS-1 and SS-3. The holes are inclined about 30° with regard to the streamwise direction. The pitch-to-diameter ratio P_h/D is 4 for all holes which gives a uniform coolant coverage on the endwall. The compound angle of the holes differs from row to row which is due to the alignment of the hole axis with the near-wall flow direction. The compound angle increases more than 20° from row PS-7a to row PS-8. The dimensionless length-to-diameter ratio of the metering section is also 75 % of the total hole length-to-diameter ratio.

8.4.1 Coolant Injection

Since less main fluid acceleration occurs at the injection location of the rows PS-7 and PS-8, only small differences in the endwall static pressures can be observed in Fig. 8.27. The distribution of the pressure coefficient C_{p2} across row PS-7b and PS-8 is found to be nearly constant for the investigated incidences. The highest static wall pressures, hence maximum C_{p2}, is determined with an average of 0.71 across row PS-7b. This is attributed to the row location farthest away from the vane throat. The pressure coefficient decreases with smaller distance from the vane throat which results into a pressure coefficient C_{p2} ranging from 0.65 to 0.71 for row PS-7a. It further drops down to 0.56 - 0.59 for row PS-8 due to its position closest to the throat location. For the holes 1-4 of row PS-7a at minimum loading, the wall pressure coefficient slightly increases from 0.65 to 0.68 with increased hole number. At larger incidence, the effect becomes slightly more pronounced showing a maximum $C_{p2} = 0.71$ for hole 4 of row PS-7a.

Due to the small changes in endwall static pressures, the coolant pressure ratio shows hardly any perceivable differences across the combined row PS-7/PS-8. This results into a uniform distribution of coolant mass flow rate across the rows PS7-PS8, as displayed in Fig. 8.28. Negligible changes are observed with variation of incidence. Small deviations in the coolant mass flow ratio across a specific row part are obtained to less than 0.2 % even for the highest incidence with an inlet flow angles of $\beta_{1,ms} = 49.8°$. Nevertheless, small differences in coolant flow rate between the separate row part are still visible which are attributed to the variation in static wall pressure between each row location. While the fraction of coolant mass flow rate that is injected through row PS-8 is about 0.5 % above the supplied mass flow rate, the mass fraction $\dot{m}_{c,i}/\dot{m}_{av}$ injected through row PS-7a lies about 0.25 - 0.3 % below it.

Figure 8.28: Distribution of coolant mass flow rate $\dot{m}_{c,i}/\dot{m}_c$ (left) and normalized hole blowing rate M_i/M_{av} (right) across rows PS-7/PS-8 with variation of loading

Since the influence of increased airfoil loading on the endwall static pressure is less severe for the combined row PS-7/PS-8, the effect on the blowing rate is also very small. The normalized blowing rate M_i/M_{av} for the injection through the combined rows PS-7/PS-8 is calculated with regard to the averaged blowing rate across all single rows. The row PS-7b holds higher values of about 1.03 to 1.09 though the absolute differences in the hole blowing rate across the row varies with incidence from 0.9 to 1.8 %. On the contrary, the blowing rate for row PS-8 is found to be 5 to 9 % less than the averaged blowing rate for all investigated rows. The differences in the blowing rate across this row are small with 0.7 up to 1.2 % compared to row PS-7b. Due to the larger pitchwise extent of row PS-7a, the increase in static pressure with larger hole number gives slightly enhanced influence on the blowing rate distribution. While the blowing rate for hole no. 1 is about 2 % less than the averaged row blowing rate, it differs for the holes 2 - 4 from 1.5 up to 3.6 % with variation of loading.

8.4.2 Adiabatic Film-Cooling Effectiveness

The film effectiveness distribution is obtained for the three incidences and an averaged row blowing rate of $M = 1.55$ and is shown on the left-hand side of Fig. 8.29 for heated CO_2 injection. Furthermore, the normalized film effectiveness is plotted along versus the non-dimensional streamwise distance s/D along two-dimensional streamlines. The computed near-wall streamlines are superimposed on the film effectiveness fields being visualized as dashed lines. The density ratio is kept at $D_R = 1.40$ with a coolant-to-main flow temperature ratio of 1.083. The momentum flux ratio is obtained to $I = 1.72$. The ratio of the momentum flux ratio divided by the area ratio between the metering and the shaped part is $I/A_R^2 = 0.092$.

The highest film effectiveness is obtained close to the film injection location of row PS-8 which is potentially attributed to the 5 % lower blowing rate and the partially overlapping film jets from row PS-7b and the outer holes 3 and 4 of row PS-7a. For the reference test case, the trajectories of the films appear to align with the hole axis direction. The results show enhanced lateral spreading closely downstream of the hole-exit region due to the diffuser effect. The separate film jets from the single rows PS-7a and PS-7b merge quickly as a result of the small spacing of 10 and 12 hole diameters, respectively. A complete layer of coolant is formed which is detaches from the endwall region at the pressure side. The films are deflected toward the suction side of the adjacent vane due to the near-wall cross flow driven by the transverse pressure gradient. As the film layer is transported downstream across the passage, the width of the film blanket is slightly reduced but still covers a large portion of about 50 % of the endwall area at the vane throat.

Nonetheless, the inner core of the film layer shows strong persistence while it is transported across the passage and the remaining dimensionless film effectiveness close to the suction side is slightly below 0.1. The layer approaches the profile wall at approximately 55 % chord. Here, the remaining coolant is up-washed along the fillet wall to be entrained into the vortical movement of the passage vortex. With superimposing the CFD-streamlines on the spatial film effectiveness, the computed near-wall flow direction agrees reasonably well with the position and orientation of the film jets footprints. With increase of loading, the cooling films experience larger deflection

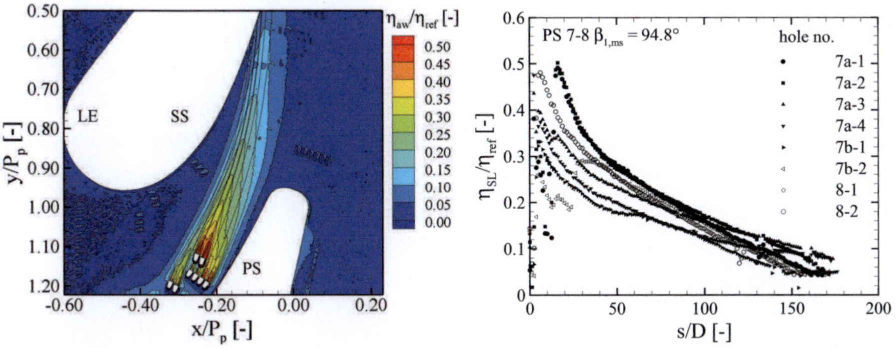

(a) $\beta_{1,ms} = 94.8°$

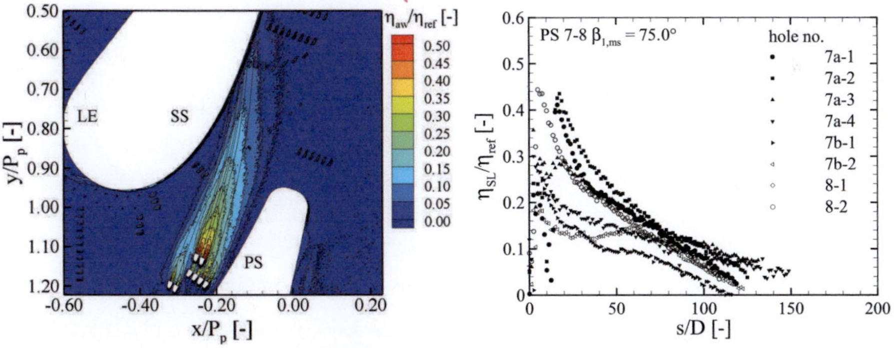

(b) $\beta_{1,ms} = 75.0°$

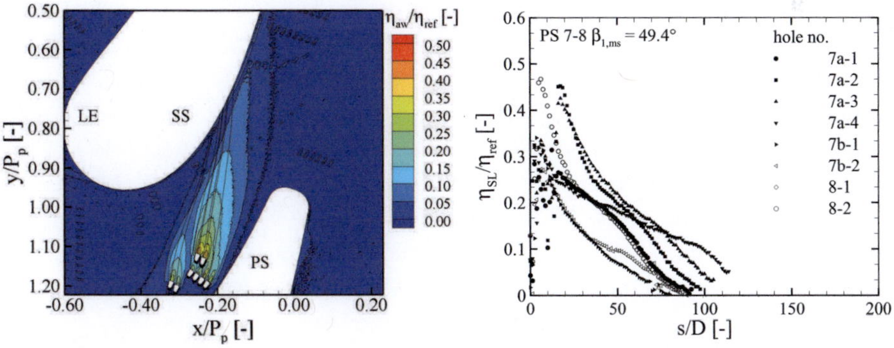

(c) $\beta_{1,ms} = 49.4°$

Figure 8.29: Normalized adiabatic film effectiveness η_{aw}/η_{ref} (left) and streamwise film effectiveness along CFD-calculated surface streamlines η_{SL}/η_{ref} (right) for row PS-7/PS-8 with variation of loading (a-c) directly downstream of the hole exits and are further turned toward the SS. This also coincides well with the enhanced curvature of the streamlines.

With higher loading, the maximum streamwise distance of the film jets is shortened which is mainly attributed to enhanced mixing with the near-wall cross flow. This leads to faster film effectiveness decay and lower magnitudes at the same streamwise position comparing different loading cases. The impingement location of the film layer on the suction side is shifted upstream toward the leading edge. It is observed at approximately 45 % for increased incidence of 15° chord which is near the injection location of row SS-7. This gives a 10 % shift with regard to the reference case at nearly zero incidence. For the case with 40° incidence and maximum loading, the position is located further upstream and closer to the leading edge at 38 % chord.

Similar findings can also be retrieved from the plots of the streamwise film effectiveness, as displayed on the right side of Fig. 8.29. For the reference case at an inlet flow angle of $\beta_{1,ms} = 94.8°$, it decreases steadily downstream of the holes until the films impinge on the profile wall at a maximum streamwise distance $s/D = 170$. The film effectiveness decays nonlinearly along the streamwise direction over the first 60 to 70 hole diameters. After that point the decrease tends to change into a linear evolution. With increasing incidence to moderate loading ($i = 15°$), the film effectiveness degrades at greater rates and the maximum streamwise distance is shortened to 130 hole diameters. For the test case with 40° incidence and maximum loading, the effects are even more pronounced. The streamwise distance is further decreased to $s/D \sim 100$ showing the fastest decay of the streamwise film effectiveness. This means a reduction in streamwise distance of about 41 % with regard to the case with minimum loading.

Considering the film effectiveness performance of row PS-7b, the hole 1 and 2 inject coolant near the lift-off line of the passage vortex. The film streak that is initiated from hole 1 is transported downstream alongside the lift-off line not crossing this border. For hole 2, the film develops quite undisturbed from the other holes of row PS-7a over the first 22 hole diameter. The streamwise film effectiveness decreases from 0.32, directly downstream of the hole exit, down to 0.2 at $s/D = 22$. Beyond this point, the holes of row PS8 add fresh coolant to the film layer at a slightly higher compound angle. Thus, the film jet from hole 2 of PS-7b partially intersect and merge at its boundary with the coolant injected from hole 1 of PS-8. Hereby a sudden rise in the streamwise film effectiveness is observed. With higher incidence, the deflection of the films of row PS-7b toward the suction surface is significantly enhanced. The trajectories are turned further away from their initial injection direction causing enhanced coolant spreading on the endwall. This also increases the distance between the films of hole 1 of PS-7b and PS-8 which results into less interaction.

The injected coolant from the first two holes of row PS-7a shows only short duration on the endwall due to the small row spacing with regard to row PS-8. High film effectiveness of 0.4 is found close to the hole exits. It decreases steeply over a short streamwise distance of $s/D = 12$ down to 0.15. At this point, the remaining coolant from the holes 3 and 4 of PS-7a merge with the injected film-cooling jets from row PS8 and the normalized film effectiveness shows a rapid increase up to 0.5. Beyond this location, the film effectiveness steadily decreases across the passage until the remaining film layer impinges on the profile suction side. Downstream of row PS-8, all individual film jets unite to form a quite compact, highly persistent coolant layer that is transported across the passage.

With higher cascade loading, the films injected through hole 3 and 4 of row PS-7a are slightly less affected at the hole exits by the enhanced cross flow than all other holes. This is attributed to their close location to the pressure surface. Further downstream, the deflection of the jets increase with enhanced decay of the streamwise film effectiveness. Finally, the row PS-8 delivers the best cooling performance which is similar for both holes. Highest normalized film effectiveness of 0.5 close to the hole exit was obtained. It steadily decreases at the same rate for both holes which results into collapsing curves. At larger incidence, the film streaks of that row experience strongly enhanced diversion away from their injection direction. That leads to faster decrease of the film effectiveness in streamwise direction but the cooling behavior is still comparable for both holes.

8.5 Multiple Row Injection at Airfoil Pressure Side

Placing sufficient cooling on the pressure side of the vane is crucial due to the jet blow off caused by the near-wall cross flow. Improved cooling performance in this area is often achieved by placing numerous staggered film hole rows along the pressure side. With minimum spacing of the rows to avoid hot spots on the surface, the individual film jets eventually interfere and merge with each other in streamwise direction. This results into differing cooling performance per row with variable injection conditions and is even more important when the profile loading is increased. To potentially reduce the amount of coolant in this endwall area, it is required to study a complete hole-group behavior and interference between individual cooling rows. Therefore, investigations with a group of film rows are conducted choosing the rows PS-4 to PS-8 due to their close location to the airfoil pressure side (see Fig. 8.30).

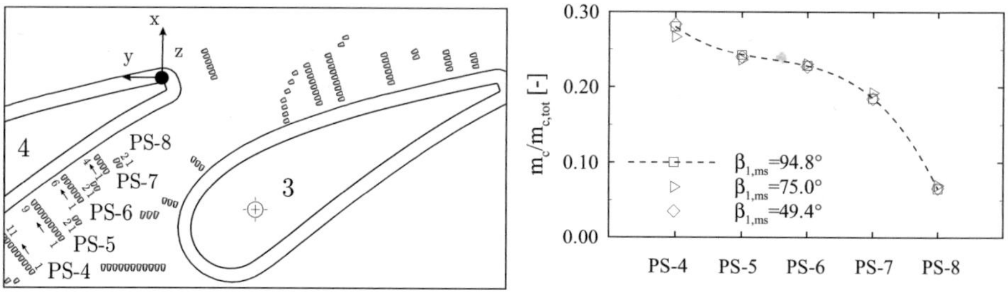

Figure 8.30: Share of coolant mass flow between the rows PS-4 to PS-8 for $\dot{m}_c/\dot{m}_m = 0.0007$ with variation of loading

Multiple row injection is established using 5 rows, denoted as PS-4 to PS-8 at the pressure side. Each of them contains a diffuser-shaped hole exit expanding over the last three hole diameter of the hole length with a moderate opening angle in lateral and in streamwise direction. All holes exhibit a shallow inclination angle of about 30° with regard to the streamwise direction. The pitch-to-diameter $P_h/D = 4$ is nearly constant for all rows. The total hole length-to-diameter ratio L_h/D for all holes is about 16. The spacing of row PS-4 and PS-5 is about 30 hole diameters increasing the streamwise distance toward the next row. The spacing is reduced to about 12 hole diameters between the outer two holes of PS-6 and PS-5. For rest of the row PS6, the holes are spaced about 25 diameters with regard to PS-5. The holes of PS-6 and the holes 1-2 of PS-7a are located farthest away from PS. Their spacing is 14 hole diameters which is increased for the holes 1-4 of PS-7b to 27 hole diameters. Each row contains a different number of holes which obviously relates to its coolant consumption. The share of coolant mass flow between the rows is displayed in Fig. 8.30. It reveals a similar coolant distribution between the film rows with variation of loading. The row PS-4, that consists of 11 holes, shows the largest mass flow fraction of about 28 % with regard to the total mass flow supplied to the film row group. It is followed by the rows PS-5 and PS-6 which have a total hole count of 9 holes. Each row consumes about 24 % of the total mass flow. The rows PS-7 and PS-8 feature less holes with 6 holes for PS-7 and only two shaped holes for row PS-8. Thus, the consumed coolant mass-flow rate decreases to about 18 % for row PS-7 and further down to fairly less than 6 % for row PS-8.

8.5.1 Coolant Injection

The main flow conditions in terms of wall pressure and mass velocity differ significantly for the combined injection through the rows PS-4 - PS-8. The injection of coolant starts at row PS-4 which is positioned closely downstream of the leading edge. Here, the wall pressure distribution is influenced by the stagnation point and the regions of accelerated fluid around the leading edge which changes distinctly with increased loading. These effects are reduced in streamwise direction. To visualize these effects, the endwall static pressure in terms of the pressure coefficient C_{p2} and the individual blowing rate is plotted in Fig. 8.31 for all cooling holes. For the reference case at $\beta_{1,ms} = 95.4°$, the static pressure deviations across the film rows is quite small. The highest pressure coefficient C_{p2} is observed for row PS-4 which rises from 0.91 to 0.95 with increasing hole number. Due to the acceleration of main fluid toward the vane throat, the wall pressures decrease in streamwise direction. The pressure gradient across the rows becomes smaller and is slightly broadened for the farther downstream film rows. Hence, the lowest static endwall pressures with $C_{p2} = 0.60$ and highest mass velocities for the main flow are obtained for row PS-8. The differences across row PS-8 are very small of less than 3 %.

Increasing the incidence leads to a shift of the stagnation region along the pressure side, as discussed in section 8.1.6. This gives changed endwall pressures across the film rows PS-4 and PS-5 which are located within the influenced wall region. For the case with moderate loading ($\beta_{1,ms} = 75.0°$), the pressure coefficient for row PS-4 increases from 0.89 at hole 1 to 0.97 at the innermost hole 11, located closest to the pressure side. When the incidence is further increased ($\beta_{1,ms} = 49.4°$), the deviations of the pressure coefficient exceed almost 16 % across row PS-4. Even row PS-8 show slightly enhanced wall pressures increasing about 6 % across the row. It is clearly visible that the lower wall pressures at the inner holes cause significantly reduced coolant pressure ratio and smaller discharge through those holes. The increase in endwall pressure results into lower mass velocities of the main flow which counterbalances the decrease of coolant mass flow rate with regard to the blowing rate. For the outer holes, the wall pressures decrease along with increasing mass velocity of the main flow. The effects become less pronounced when proceeding further downstream to greater row number.

Considering the blowing rate distribution, the averaged row blowing rate decreases with larger row number due to the increasing main flow mass velocities. The largest blowing rate is obtained for row PS-4 and decreases continually for all following rows. For the reference case ($\beta_{1,ms} = 95.4°$), the distribution of the hole blowing rate shows less deviation compared to

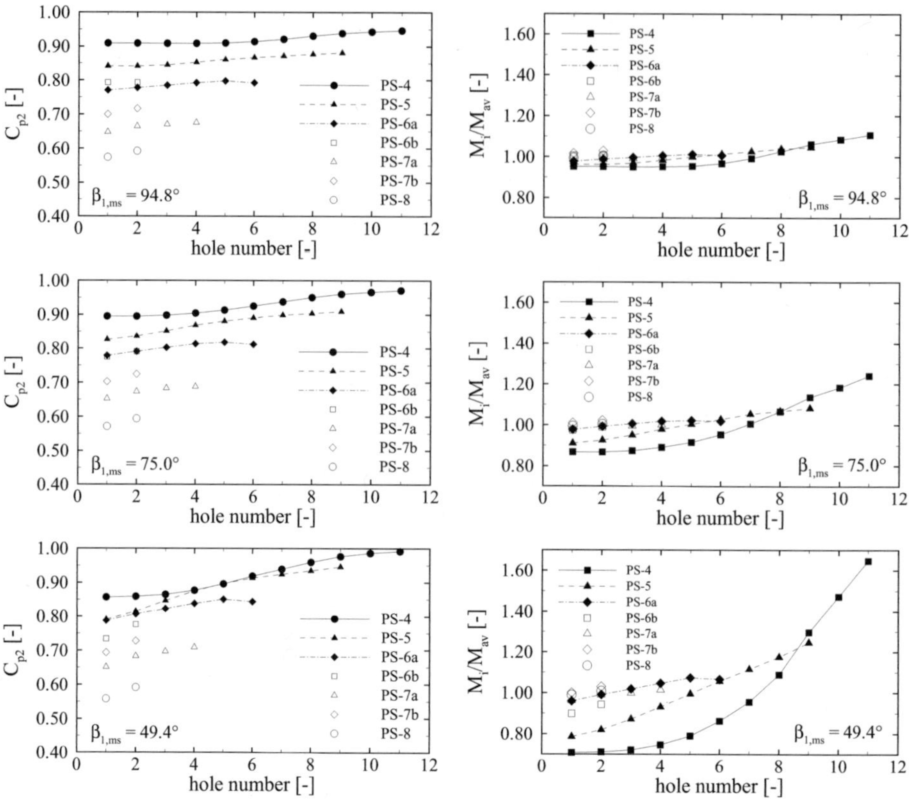

Figure 8.31: Pressure coefficient C_{p2} (left) and normalized hole blowing rate M_i/M_{av} for multiple row injection using the film rows PS-4 to PS-8 with variation of loading

the test cases with increased incidence. The loading effect is mainly concentrated on the rows PS-4 and PS-5. Asymmetric evolution of the hole blowing rate can be seen for row PS-4 with the holes 1-5 delivering nearly similar blowing rates. This is mainly attributed to the constant endwall pressure for these holes with deviations of 4 % below the row blowing rate. At hole 6, the blowing rate starts to rise up to 10 % above the row blowing rate. With the next cooling row PS-5, the variation of the blowing rate is a little less pronounced which yields ± 4 % deviation across the row. Minimum differences in the blowing rate are observed for the row PS-8 with less than 1 %.

Increasing the profile incidence, the differences in the local blowing rate are enhanced due to relocation of the stagnation point and varying mass velocity of the main flow. For row PS-4 at moderate incidence ($\beta_{1,ms} = 75.0°$), the blowing rate for the first hole is lower about 13 % than the row blowing rate. Nonetheless, it increases up to 23 % for the last hole. The differences in the blowing rate for row PS-5 are slightly less pronounced. It shows deviations of the hole blowing rate of ± 8 % across the film row. For the other rows PS-6 to PS-8, the differences are observed to be about ± 2 % or less.

At maximum loading ($\beta_{1,ms} = 49.4°$), the effect of the relocating stagnation region on the endwall pressure field and the blowing rate distribution is further spread in streamwise direction. In this case, the row PS-4 exhibits the largest variation in blowing rate from 30 % at hole 1 up to 60 % above the averaged row blowing rate for the last hole 11. Due to the moderate streamwise spacing between row PS-4 and PS-5, an increased influence is recognizable on the blowing rate distribution for row PS-5. Differences are observed within a range of ± 20 % with regard to the averaged row blowing rate. Even for the row PS-6a which is located more than 40 hole diameters downstream of row PS-4, the effect can still be observed. Although differences in the hole blowing rate are comparatively small in the range of ± 1 % for zero incidence, the deviations become significantly enhanced with increased loading, showing a gradient of up to ± 5 % across the row. For the other rows PS-7 and PS-8, the variation at maximum loading is less than ± 3 %.

8.5.2 Adiabatic Film-Cooling Effectiveness

Film effectiveness distributions are obtained for multiple row injection with the film rows PS-4 to PS-8 for different coolant-to-main flow mass fractions within $\dot{m}_c/\dot{m}_m = 0.0007 \ldots 0.0015$. The results are discussed for the smallest mass fraction of $\dot{m}_c/\dot{m}_m = 0.0007$ which is equivalent of an averaged group blowing rate of $M = 1.50$. A coolant-to-main flow density ratio of 1.41 was established using heated CO_2 which equals to a temperature ratio of $T_R = 1.10$. Similarly to the case of single row injection, three different loading conditions were investigated. The spatial adiabatic film effectiveness and the streamwise development of the laterally-averaged film effectiveness are presented in Fig. 8.32 on the left-hand side. Near-wall CFD- streamlines are superimposed onto the film effectiveness fields for visualizing the flow field close to the endwall. Furthermore, the film effectiveness data is point-wise averaged in pitchwise direction for each streamwise position over all streamlines of each row. The so derived laterally-averaged film effectiveness is then plotted versus the normalized streamwise distance s/D, as displayed on the right-hand side of Fig. 8.32.

The holes of the combined rows PS-4 to PS-8 are located downstream of the lift-off line of the passage vortex. Hence, these rows inject coolant into the distorted three-dimensional endwall boundary layer which is mostly affected by the cross flow flowing from the pressure to the suction side. Fig. 8.32 clearly shows that the jet trajectories agree well the viscid surface streamlines computed by the CFD-simulations. The hole axis for the rows PS-4 to PS-7a are aligned nearly tangential to the pressure side contour to map the local near-wall flow direction. For the other holes of row PS-7b and PS-8, the compound angle is slightly increased to account for the stronger streamline's curvature. Near the hole exits, the trajectories are matched with the flow direction imposed by the streamlines guaranteeing sufficient coolant coverage. With increasing streamwise distance, the film-cooling jets are diverted away from the pressure surface by the near-wall cross flow which leaves triangularly-shaped, uncooled gaps in the rear part of the vane endwall corner.

Since the row PS-4 experiences the highest static pressures and has the largest number of holes, it receives the highest mass flow compared to the other rows. Combined with the low mass velocities of the main flow that occur at its hole exits, the blowing rate is highest. Observing the film effectiveness distributions, the row PS-4 shows poor cooling performance resulting into lowest film effectiveness compared to all other rows. Accordingly it swiftly decays in streamwise direction over 30 hole diameter toward row PS-5. The remaining film effectiveness is less than 15 % at the injection locations of PS-5. With placing additional rows further downstream, the mass velocity of the main flow increases which results into decreased row blowing rates. The closer the hole exits are located toward the vane throat the lower the blowing rate. This decreases the coolant momentum and results into higher film effectiveness. Considering row PS-8, the blowing rate is slightly below one and the normalized film effectiveness η_{aw}/η_{ref} that is obtained directly downstream of the hole exits is significantly increased above 0.7.

Due to the moderate spacing of the rows, the film layer emerged from each row is not completely consumed in streamwise direction. The film-cooling jets tend to partially or fully superimpose and merge with the injected coolant from the next row. However, the resulting film layer shows surprisingly large spread across the endwall and distinct persistence. At the vane throat, it covers approximately 40 % of the endwall area. After releasing the last films from row PS8, the layer is diverted away toward the suction side. Due to its high persistence and the large amount of injected coolant through the complete hole group, film effectiveness levels of about 0.15 - 0.20 can still be observed when the films approach the suction side surface of the adjacent vane. At this point, the entire film layer impinges onto the profile wall at about 58 % chord, is blown off the endwall and entrained into the vortical movement of the passage vortex. With higher loading, the streamwise decay of film effectiveness is found to be enhanced and the impingement region is shifted in upstream direction. Considering the test case with maximum loading ($\beta_{1,ms} = 49.4°$), the film effectiveness drops down rapidly to 0.08 at the point where the film layer approaches the suction side surface. That indicates intensified mixing and interaction of the coolant with the highly-turbulent, vortical motion of the near-wall cross flow. Due to enhanced diversion of the film layer away from the pressure side, the impingement region is shifted upstream to 40 % chord.

The enhanced degradation of the film effectiveness with increased loading is also visible in the evolution of the laterally-averaged streamwise film effectiveness, as plotted in Fig. 8.32. For the rows PS-4 to PS-6a at minimum loading, a superposition of the injected films from upstream rows was observed. This leads to alternate decline and rise of the streamwise film

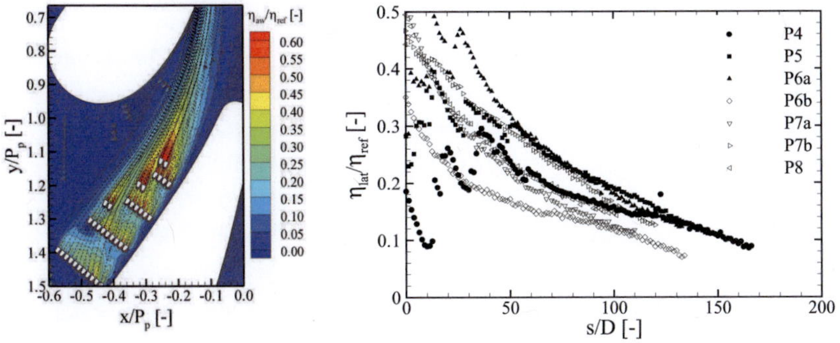

(a) $\beta_{1,ms} = 94.8°$

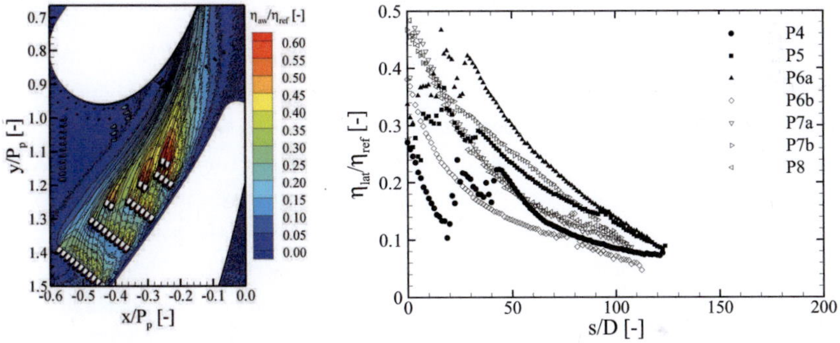

(b) $\beta_{1,ms} = 75.0°$

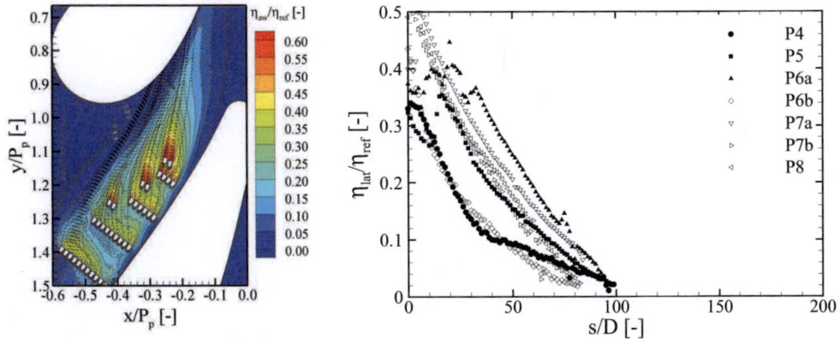

(c) $\beta_{1,ms} = 49.4°$

Figure 8.32: Normalized adiabatic film effectiveness (left) and laterally-averaged streamwise film effectiveness (right) for CO_2-injection through rows PS-4 to PS-8 with $\dot{m}_c/\dot{m}_m = 0.0007$ and elevated coolant density ratio $D_R = 1.41$ with variation of loading (a-c) effectiveness when the cooling films approach the next injection location. This effect is reduced with increased incidence and vanishes almost completely for the case with maximum loading, due to the separation of individual jets with more pronounced diversion. For the reference case ($\beta_{1,ms}$ = 95.4°), the film effectiveness shows an exponential decay until a streamwise distance of $s/D = 70$. It then changes to a nearly linear behavior dropping down to values of η_{lat}/η_{ref} = 0.1 for the impingement region at $s/D = 150$. The highest film effectiveness is obtained downstream of row PS-8 above 0.5. With increase of the incidence ($\beta_{1,ms}$ = 75.0°), the film effectiveness degrades more swiftly over a decreased streamwise distance of $s/D = 120$. At maximum loading ($\beta_{1,ms}$ = 49.4°), the decay of the film effectiveness pass even more swiftly with a near-linear streamwise evolution over less than 100 hole diameters.

With increase of airfoil incidence, the lift-off line of the passage vortex is moved upstream detaching earlier from the pressure surface. Additionally, the stagnation region shifts in streamwise direction along the pressure side. This causes the pressure gradient to be redirected more in the lateral direction. Likewise, the film trajectories show larger deflection away from their initial injection direction toward the suction side. The enhanced jet diversion at the hole exits cause the coolant inside the hole to alternately separate from the diffuser sidewalls. This drives accumulated coolant concentration in isolated endwall areas and therefore leads to locally elevated film effectiveness. This effect can be observed for row PS-4 at the test case at maximum loading ($\beta_{1,ms}$ = 49.4°). Here, strongly deflected film trajectories downstream of holes 8 - 11 lead to coolant transport outwards toward the lift-off line of the passage vortex. Hence, improved coolant concentration at the outer holes 1-4 are observed and the film effectiveness increases about 12 % compared to the reference case at minimum loading. For the endwall region upstream of the holes 6-9 of row PS-5, it naturally results into reduced film effectiveness due to the absence of cooling.

9 Conclusions

Experimental investigations were performed on the platform cooling configuration of a highly-loaded nozzle guide vane focusing on the coolant behavior on the endwall under increased airfoil loading. The cascade flow field was extensively analyzed using pneumatic probes, wall taps, CTA and surface flow visualizations. The determination of the film effectiveness distributions on the film-cooled endwall was carried out by temperature-sensitive paints. Although a low thermal conductivity material was used for the cooled cascade endwall, the wall temperatures still showed influences from three-dimensional heat conduction within the test plate. Therefore, detailed solid-state thermal analysis were performed based on a finite-element method. The computed conductive heat flux pattern was integrated into the data reduction procedure to correct the measurements. The FE-simulations were able to capture the major conduction pattern visible upstream and around the hole breakouts. Besides this, increased wall temperatures were observed at the film layer edges and between the individual film-cooling jets induced by the hot core of the film jets which initiates conductive heat fluxes toward uncooled wall regions. This effect artificially increases the wall temperature and consequently the film effectiveness. Assessing the streamwise development of the film effectiveness, near-wall streamlines derived from CFD-simulations were superimposed onto the surface temperature fields. They were calculated at a spanwise plane 0.7 %, with regard to profile midspan, distant from the endwall. To select the appropriate wall distance, the streamline pattern was compared to oil-film flow visualizations and the measured temperature footprints of the cooling films.

The scope of the experimental work was subdivided into two parts. First, flat plate film-cooling investigations were conducted with a cylindrical and a diffuser hole geometry for validation purposes of the TSP-technique. The obtained film effectiveness for the cylindrical holes clearly show significant influence of the density ratio. The comparison of the centerline film effectiveness with previous studies shows good agreement for the case with low coolant momentum. As the coolant momentum increases, slightly increased differences are observed which are mainly attributed to different fluid and geometrical boundary conditions. Compared to the cylindrical holes, the experimental results of the fan-shaped holes reveal superior cooling performance with higher film effectiveness and improved wall coverage. The influence of the density ratio on the coolant blow-off is negligible small for all blowing rates. The centerline film-cooling effectiveness reveals good qualitative and quantitative agreement with the previous studies. Laterally-averaged film effectiveness is slightly higher than the data from the previous studies which may be attributed to lateral conduction effects induced by the hot films.

The second and main part of the experimental program was carried out on the airfoil cascade with endwall cooling. Prior to the endwall cooling experiments, the aerodynamic performance of the airfoil and the flow field at the cascade exit was examined. The loss characteristic shows a strong insensitivity of the profile loss over a range of $\pm 10°$ incidence with less than 5 % change. With increased incidence up to + 40.6°, the cascade turning, hence the airfoil loading rises. An increase of about 200 % for the area-averaged secondary loss coefficient was observed at maximum loading. The results of the total pressure loss show more pronounced loss peaks paired with larger lateral and spanwise extent. Moreover, an enhancement of the secondary flow vectors and increased streamwise vorticity were observed with increased loading. The airfoil pressure distribution reveals a front-loaded profile with symmetric positions of the PS and SS maximum near the leading edge. At increased loading, the maximum pressure difference is increased about 21 % and shifts downstream along PS. Based on the observed results, three loading cases were specified with the reference test case near zero incidence. This is close to the nominal operating point of the real-engine vane. The other two test cases were done at 15° and 40.4° incidence with the latter considered as the case of maximum loading. Furthermore, the experimental results of the cascade flow field are compared to the numerical simulations. Good qualitative and quantitative agreement is found for the cascade flow field and the endwall pressure distribution. Differences are observed for the total pressure losses in which the numerical simulations always predict higher losses. The maximum deviation of the endwall loss coefficient between the experiments and CFD is quantified with 42 %. It decreases with larger incidence down to 17 %.

In a general context, the experiments have shown that the variation of loading leads to significant influence on the platform cooling due to the enhancement of the secondary flow. If such large variation of the inlet flow angle shall be considered for the cooling design of the first vane, these effects have to be integrated into the design system to avoid hot spots and enlarged, uncooled endwall areas. The streamwise film effectiveness evaluated for the investigated cooling holes can be used to develop correlations as function of loading and blowing parameters. Film-cooling injection was established for three single rows placed at the cascade inlet (SS-1), at the profile SS (SS-3) and in the region near the vane throat (PS7-PS8) at the PS. Furthermore, multiple hole injection with 5 rows near the pressure side was done. The results showed that cooling injection, which is placed in the front part of the passage, is dominated by the near-wall cross flow and the secondary flow vortices. The rear-part of the platform is less influenced. At the cascade inlet (row SS-1), the wall pressures are significantly influenced by the potential flow effect of the leading edge. As a result of this, the hole blowing rate shows -4 up to +7 % deviation for minimum loading. This effect is enhanced to -6 and +11 % for maximum loading. Along the pressure and suction side up to 30 % chord, the wall pressure field is mainly influenced by the accelerated flow around the leading edge. The effect is enhanced when the airfoil loading is increased and the stagnation region is shifted downstream along the PS. At the suction side close to the leading edge (row SS-3), the variation of the local blowing rate is about ± 8 % for the operating point of the vane and increases up to $\pm 14\,\%$ for maximum vane loading. For hole locations further downstream these influences are visibly reduced. Considering the injection locations of row PS7-PS8, the deviation of the hole blowing rate is negligible with less than $\pm 2\,\%$.

Adiabatic film effectiveness marks a significant influence of the near-wall cross flow on the coolant injection and its streamwise development. Close to the injection locations, the film-cooling jets were deflected from their initial trajectories dictated by the hole axis. The coolant is transported in streamwise direction over long distances until reaching the suction side of the adjacent vane where the film layer impinges onto the airfoil wall. Here the coolant is completely consumed loosing its cooling capacities. It is then up-washed along the fillet surface and entrained into the the vortical movement of the secondary flow. At the inlet region, additional influences by the horse-shoe vortex were observed on the coolant distribution even at minimum vane loading. Therein the vortical structure of its branches acts as barrier for the injected films along the endwall separation lines, preventing them from crossing this border. As a result of this, the endwall region directly upstream of the leading edge is characterized by very low film effectiveness. These effects were significantly intensified with increased loading.

Less deflection of the film layer at minimum loading was observed for rows close to the pressure side. Here the persistence of the injected film blanket is strong enough to transport cooling fluid across the entire airfoil passage. Remarkably, the film effectiveness still showed levels of 0.1 when approaching the impingement location at the profile wall. This persistence is slightly reduced with increased loading due to intensified mixing and interaction with the main gas path. At maximum loading, the films detach from the pressure side contour leaving an enlarged, uncooled endwall area near the profile. For multiple row injection, a superposition of the injected films from upstream rows was observed. This leads to alternate decline and rise of the streamwise film effectiveness when the cooling films approach the next injection location. This behavior is considerably dependent on the cascade loading and its influence of near-wall cross flow pattern on the coolant trajectories. With increased loading, the diversion of the film-cooling jets is enhanced which reduces the endwall coverage with coolant. Additionally, this effect cause an upstream shift of the impingement region of about 10 to $18\,\%$, depending on the investigated film row. Significantly enhanced mixing of the coolant resulted in a more swift streamwise decay of the film effectiveness with increased loading. A reduction in streamwise distance ranging from 33 up to $50\,\%$ was observed which equals 40 hole diameters (row SS-3) up to 100 hole diameters (row SS-1).

For further investigations, it is suggested to extend the correction method by incorporating the jet-induced conduction effects using a coupled solid-state thermal analysis with CFD-simulations of the flow field. Also, further measurements of the endwall heat transfer with coolant injection and variation of loading should be considered to assess the overall cooling performance of these platform holes.

References

[1] ANDREOPOULOS, J.: *On the Structure of Jets in a Cross flow.* Journal of Fluid Mechanics, Cambridge University Press, 157:163–197, 1985.

[2] ANDREOPOULOS, J.; RODI, W.: *Experimental Investigation of Jets in a Crossflow.* Journal of Fluid Mechanics, Cambridge University Press, 138:93–127, 1984.

[3] AUF DEM KAMPE, T.; VÖLKER S.: *Numerical Investigation of Flat Plate Film Cooling and Analysis of Variance of Relevant Flow Parameters.* ETC, European Turbomachinery Conference, Graz, 2009.

[4] BALDAUF, S.: *Filmkühlung thermisch hochbelasteter Oberflächen: Korrelation thermographischer Messungen.* Ph.D. thesis, Universität Karlsruhe, Universität Karlsruhe, Institut für Thermische Strömungsmaschinen, 2001.

[5] BALDAUF,S. ; SCHULZ, A.; WITTIG, S.: *High Resolution Measurements of Local Effectiveness by Discrete Hole Film Cooling.* ASME J. Turbomach., 123:758–765, 2001.

[6] BARIGOZZI G.; BENZONI, G.; FRANCHINI, G.; PERDICHIZZI, A.: *Fan-shaped Hole Effects on the Aero-Thermal Performance of a Film Cooled Endwall.* ASME J. Turbomach., 128:43–25, 2006.

[7] BARIGOZZI G.; FRANCHINI, G.; PERDICHIZZI, A.: *Inlet Turbulence Intensity Effects on Secondary Flows and Adiabatic Effectiveness of a Film Cooled Endwall.* Proceedings of the 7th European Conference on Turbomachinery Fluid Dynamics and Thermodynamics,, Athens, Greece:pp. 1105–1116, 2007.

[8] BELL, J. H.; SCHAIRER, E. T.; HAND, L. A.; MEHTA, R.: *Surface Pressure Measurements using Luminescent Coatings.* Annu. Rev. Fluid Mech., 33:155–206, 2001.

[9] BEVINGTON, P.R.; ROBINSON, D.K.: *Data Reduction and Error Analysis for the Physical Sciences.* 3rd Ed., McGraw Hill, New York, 2002.

[10] BLAIR, M. F.: *An Experimental Study Of Heat Transfer And Film-Cooling On Large-Scale Turbine Endwalls.* ASME J. Heat Transfer, 96:524–529, 1974.

[11] BLANCHARD, J. N.; BRUNET, Y.; MERLEN, A.: *Influence of a Counter Rotating Vortex Pair on the Stability of a Jet in a Cross Flow: An Experimental Study by Flow Visualizations.* Experiments in Fluids, 26:63–74, 1999.

[12] BOGARD, D. G.; THOLE, K. A.: *Gas Turbine Film Cooling.* AIAA, Journal of Propulsion and Power, 22:249–270, 2006.

[13] Bons, J. P.; MacArthur, C. D.; Rivir, R. B. : *The Effect of High Free-Stream Turbulence on Film Cooling Effectiveness.* ASME J. Turbomach, 118:814–824, 1996.

[14] Boussinesq, J.: *Theorie de l'ecoulement tourbillant.* Presentes par Divers Savants Acad. Sci. Inst. Fr., 23:46–50, 1877.

[15] Boyle, R. J.; Russell, L. M.: *Experimental Determination Of Stator Endwall Heat Transfer.* ASME J. Turbomach, 112:547–558, 1990.

[16] Brauckmann, D.: *Experimentelle Untersuchung von Filmkühlungsvorgängen an konturierten Bohrungen.* Ph.D. thesis, Universität Stuttgart, Universität Stuttgart, Institute of Aerospace Thermodynamcis, 2006.

[17] Brauckmann, D.; von Wolfersdorf, J.: *Infrared Thermography with In-Situ Calibration using Thermochromic Liquid Crystals Applied to Film Cooling.* ASME Paper GT-2004-53855, 2004.

[18] Brauckmann, D.; von Wolfersdorf, J.: *Influence of Compound Angle On Adiabatic Film Cooling Effectiveness And Heat Transfer Coefficient For a Row Of Fanshaped Film Cooling Holes.* ASME Paper GT-2005-68036, 2005.

[19] Brittingham, R. A.; Leylek, J. H.: *A Detailed Analysis of Film-Cooling Physics: Part IV - Compound-Angle Injection With Shaped Holes.* ASME J. Turbomach, 122:133–145, 2000.

[20] Bunker, R. S.: *Film Cooling Effectiveness Due to Discrete Holes within a Transverse Surface Slot.* ASME Paper, GT-2002-30178, 2002.

[21] Bunker, R. S.: *A Review of Shaped Hole Turbine Film-Cooling Technology.* ASME J. Heat Transfer, 127:441–453, 2005.

[22] Burd, S.; Simon, T.: *The Influence of Coolant Supply Geometry on Film Cooling Exit Flow and Surface Adiabatic Effectiveness.* ASME Paper, 97-GT-026, 1997.

[23] Burd, S. W.; Kaszeta, R.W.; Simon, T.W.: *Measurements in Film Cooling Flows: Hole L/D and Turbulence Intensity Effects.* ASME J. Turbomach., 120:791–797, 1996.

[24] Campbell, B. T.; Liu, T.; Sullivan, J. P.: *Temperature Sensitive Fluorescent Paint Systems.* AIAA Aerospace Ground Testing Conference, Colorado Springs CO, 94-2483, 1994.

[25] Cattafesta, L. N.; Liu, T.; Sullivan, J. P.: *Uncertainty Estimates for Temperature Sensitive Paint Measurements with CCD camera.* AIAA, 36 (11):2102–2108, 1998.

[26] Charbonnier, D.; Ott, P.; Jonsson, M.; Cottier, F.; Köbke, Th.: *Experimental and Numerical Study of the Thermal Performance of a Film-Cooled Turbine Platform.* ASME Paper, GT2009-60306, 2009.

[27] Cortelezzi, L.; Karagozian, A. R.: *On the Formation of the Counter-Rotating Vortex Pair in Tranverse Jets.* Journal of Fluid Mechanics, Cambridge University Press,

446:347–373, 2001.

[28] CRAFTON, J.W.: *The Impingement of Sonic and Sub-Sonic Jets onto a Flat Plate at Inclined Angles.* Ph.D. thesis, Purdue University, Purdue University, Dept. of Mechanical Engineering, 2004.

[29] CUTBIRD, J. M.; BOGARD, D. G.: *Effects of Coolant Density Ratio on Film Cooling Performance on a vane.* ASME Paper GT-2003-38582, 2003.

[30] DUDEN, A.: *Strömungsbeeinflussung zur Reduzierung der Sekundärströmungen in Turbinengittern.* Ph.D. thesis, Universität der Bundeswehr München, UniBw München, Fakultät für Luft- und Raumfahrttechnik, 1999.

[31] EKKAD, S. V.; ZAPATA, D.; HAN, J. C.: *Film Effectiveness Over a Flat Surface With Air and CO_2 Injection Through Compound Angle Holes Using a Transient Liquid Crystal Image Method.* ASME J. Turbomach., 119:580–586, 1997a.

[32] EKKAD, S. V.; ZAPATA, D.; HAN, J. C.: *Heat Transfer Coefficients Over a Flat Surface With Air and CO_2 Injection Through Compound Angle Holes Using a Transient Liquid Crystal Image Method.* ASME J. Turbomach., 119:587–593, 1997b.

[33] FOSTER, N. W.; LAMPARD, D.: *The Flow and Film Cooling Effectiveness Following Injection Through a Row of Holes.* ASME J. Eng. Power, 102:584–588, 1980.

[34] FRIC, T. F.; ROSHKO, A.: *Structure in the Near-Field of the Transverse Jet.* Proc. of 7th Symp on Turbulent Shear Flow, pages 6.4.1–6.4.6, 1989.

[35] FRIC, T. F.; ROSHKO, A.: *Vortical Structure in the Wake of a Transverse Jet.* Journal of Fluid Mechanics, Cambridge University Press, 279:1–47, 1994.

[36] FRIEDRICHS, S.: *Endwall Film-Cooling in Axial Flow Turbines.* Ph.D. thesis, Whittle Laboratory, Cambridge University, Engineering Department, 1997.

[37] FRIEDRICHS, S.; HODSON, H. P.; DAWES, W. N.: *Design of an Improved Endwall-Film-Cooling Configuration.* ASME J. Turbomach, 121:772–780, 1996.

[38] FRIEDRICHS, S.; HODSON, H. P.; DAWES, W. N.: *Distribution of Film-Cooling Effectiveness On A Turbine Endwall Measured Using The Ammonia And Diazo Technique.* ASME J. Turbomach, 119:613–621, 1996.

[39] FRIEDRICHS, S.; HODSON, H. P.; DAWES, W. N.: *Aerodynamic Aspects Of Endwall Film-Cooling.* ASME J. Turbomach, 119:786–793, 1997.

[40] GAUGLER, R. E.; RUSSELL, L.M: *Comparsion of Visualized Turbine Endwall Secondary Flows And Measured Heat Transfer Patterns.* ASME J. Eng. Gas Turbines Power, 106:168–172, 1984.

[41] GOLDSTEIN, R. J.; CHIANG, H. D.; ELOVIC, E.: *Effect of Surface Roughness on Film Cooling Performance.* ASME J. Eng. Gas Turbines Power, 107:111–116, 1985.

[42] GOLDSTEIN, R. J.; ECKERT, E. R. G.; BURGGRAF, F.: *Effects of Hole Geometry and*

Density on Three-Dimensional Film Cooling. Int. Journal of Mass and Heat Transfer, 17:595–607, 1974.

[43] GOLDSTEIN, R. J.; SPORES, R. A.: *Turbulent Transport on the Endwall in the Region Between Adjacent Turbine Blades.* ASME J. Heat Transfer, 110:862–869, 1988.

[44] GOLDSTEIN, R.J.; SPORES, R.A.: *Turbulent Transport on the Endwall in the Region Between Adjacent Turbine Blades.* ASME J. Heat Transfer, vol. 110:pp. 862–869, 1988.

[45] GRAZIANI, R. A.; BLAIR, M. F.; TAYLOR, J. R.; MAYLE, R. E.: *An Experimental Study of Endwall and Airfoil Surface Heat Transfer in a Large-Scale Turbine Blade Cascade.* ASME J. Eng. Gas Turbines Power, 102:257–267, 1980.

[46] GREGORY-SMITH, D.G.; GRAVES, C. P.; WALSH, J.A.: *Growth of Secondary Losses and Vorticity in an Axial Turbine Cascade.* ASME Paper, 87-GT-114, 1987.

[47] GRITSCH, M.: *Experimentelle Untersuchungen zum aerothermischen Verhalten nicht-zylindrischer Filmkühlbohrungen.* Ph.D. thesis, Universität Karlsruhe, Universität Karlsruhe, Institut für Thermische Strömungsmaschinen, 1998.

[48] GRITSCH,M. ; SCHULZ, A.; WITTIG, S.: *Adiabatic Wall Effectiveness Measurements of Film-Cooling Holes with Expanded Exits.* ASME J. Turbomach., 120:549–556, 1998.

[49] HAMMER, M. P.: *Using Temperature Sensitive Paint.* AIAA paper, Aerosp. Sci. Meet. Exhib., 40th Reno NV, 2002-0742, 2002.

[50] HAN, J. C., DUTTA, S., EKKAD, S. V.: *Gas Turbine Heat Transfer and Cooling Technology.* Taylor and Francis, London, 2000.

[51] HARAGASMA, S. P.; BURTON, C. D.: *Film-Cooling Research On The Endwall Of A Turbine Nozzle Guide Vane In A Short Duration Annular Cascade: Part I - Experimental Technique and Results.* ASME J. Turbomach, 114:734–740, 1992.

[52] HARRISON, K. L.; BOGARD, D. G.: *CFD-Predictions of Film Cooling Adiabatic Effectivness for Cylindrical Holes Embedded in Narrow and Wide Transverse Trenches.* ASME Paper, GT2007-28005, 2007.

[53] HAWTHORNE, W.R.: *Rotational Flow Through Cascades.* J. Mech. & Appl. Math., vol. 3, 1955.

[54] HAY, N.; LAMPARD, D.: *The Discharge Coefficient of flared film cooling holes.* ASME Paper, 95-GT-15, 1995.

[55] HENEKA C.; SCHULZ, A.; BAUER, H. J.: *Effect of Internal Rib Configurations on the Discharge Coefficient of a 30-Degree Inclined Film Cooling Hole.* ASME J. Heat Transfer, 40 no. 10, 2010.

[56] HUBNER, J. P.; CAROLL, B. F.; SCHANZE, K. S.: *Heat Transfer Measurements in Hypersonic Flows using Luminescent Coating Techniques.* Journal of Thermophysics and Heat Transfer, 16 (4):516–522, 2002.

[57] Hyams, D. G.; Leylek, J. H.: *A Detailed Analysis of Film-Cooling Physics: Part III - Streamwise Injection With Shaped Holes.* ASME J. Turbomach, 122:122–132, 2000.

[58] Jabbari, M. Y.; Marston, K. C.; Eckert, E. R. G.; Goldstein, R. J.: *Film-Cooling of the Gas Turbine Endwall By Discrete-Hole Injection.* ASME J. Turbomach, 118:278–284, 1996.

[59] Jones, T. V.: *Theory for the use of foreign gas in simulating film cooling.* Int. J. Heat and Fluid Flow, 20:349–354, 1999.

[60] Jumper, G. W.; Elrod, W. C.; Rivir, R. B.: *Film Cooling Effectiveness in High Turbulence Flow.* ASME J. Turbomach., 113:479–483, 1991.

[61] Kadotani, K.; Goldstein, R. J.: *Effect of Mainstream Variables on Jets Issuing from a Row of Inclined Holes.* ASME J. Turbomach., 101:298–304, 1979.

[62] Kadotani, K.; Goldstein, R. J.: *On the Nature of Jets Entering a Turbulent Flow Part A - Jet Mainstream Interaction.* ASME J. Eng. Power, 101:459–465, 1979.

[63] Kadotani, K.; Goldstein, R. J.: *On the Nature of Jets Entering a Turbulent Flow Part B - Film Cooling Performance.* ASME J. Eng. Power, 101:466–470, 1979.

[64] Kamotani, Y.; Greber, I.: *Experiments on a Turbulent Jet in a Cross Flow.* AIAA Journal, 10:1425–1429, 1972.

[65] Kang, M.; Kohli, A.; Thole, K. A.: *Adiabatic Effectiveness Measurements Of Endwall Film-Cooling For A First Stage Vane.* ASME Paper GT-2004-53326, 2004.

[66] Kaszeta, R. W.: *Measurements in Film Cooling Flows with Lateral Injection.* Ph.D. thesis, University of Minnesota, University of Minnesota, Heat Transfer Laboratory, Minneapolis, MN 55455, USA, 1998.

[67] Kays, W.; Crawford, M.; Weigand, B.: *Convective Heat and Mass Transfer.* McGraw Hill Book Co., New York, 2004.

[68] Kelso, R. M.; Lim, T. T.; Perry, A. E.: *An Experimental Study of Round Jets in Cross-Flow.* Journal of Fluid Mechanics, Cambridge University Press, 306:111–144, 1996.

[69] Kissel, H.: *Filmkühlung bei komplexen Innenströmungen.* Ph.D. thesis, Universität Stuttgart, Universität Stuttgart, Institute of Aerospace Thermodynamcis, 2007.

[70] Knost, D. G.; Thole, K. A.: *Heat Transfer and Flowfield Measurements in the Leading Edge Region of a Stator Vane Endwall.* ASME J. Turbomach, 121:558–568, 1999.

[71] Kohli, A.; Bogard, D. G.: *Improving Film Cooling Performance Using Airfoil Contouring.* ASME Paper, GT2006-90852, 2006.

[72] Kohli, A.; Bogard, D.G.: *Measurements in Film Cooling Flows: Hole L/D and Turbulence Intensity Effects.* ASME J. Turbomach., 120:791–797, 1996.

[73] Kost, F.; Nicklas, M.: *Film-Cooled Turbine Endwall In A Transonic Flow Field: Part 1 - Aerodynamic Measurements.* ASME J. Turbomach, 123:709–719, 2001.

[74] Kunze, M.; Vogeler, K. : *LDA-measurement in a Film-Cooling Hole With Cross Flow.* Technical Report, 2003.

[75] Langston, L.S.: *Crossflows in a Turbine Cascade Passage.* ASME J. Eng. Power, vol. 102:pp. 866–874, 1980.

[76] Langston, L.S.: *Secondary Flows in Axial Turbines: A Review.* Annals of the New York Academy of Sciences, vol. 934:pp. 11–26, 2006.

[77] Lee, S. W.; Park, S. W.; Lee, J. S.: *Flow Characteristics Inside Circular Injection Holes Normally Oriented To a Cross Flow: Part I Flow Visualizations and Flow Data in the Symmetry Plane.* ASME Paper, 2000-GT-256, 2000.

[78] Lemmon E. W.; Jacobson, R. T.: *Viscosity and Thermal Conductivity Equations for Nitrogen, Oxygen, Argon and Air.* International Journal of Thermophysics, 25 no.1:21–69, 2004.

[79] Liess, C.: *Experimental Investigation of Film Cooling with Ejection from a Row of Holes for the Application to Gas Turbines.* ASME J. Eng. Power, 97:21–27, 1975.

[80] Ligrani, P. M.; Ciriello, S.; Bishop, D. T.: *Heat Transfer, Adiabatic Effectiveness and Injectant Distributions Downstream of a Single Row and Two Staggered Rows of Compound Angle Film-Cooling Holes.* ASME J. Turbomach, 114:687–701, 1992.

[81] Liu, T.; Campbell, B. T.; Bruns, S.; Sullivan, J. P.: *Temperature- and Pressure Sensitive Luminscent Paints in Aerodynamics.* Applied Mechanics Reviews, 50 (4):227–246, 1997.

[82] Liu, T.; Campbell, B. T.; Sullivan, J. P.: *Remote Surface Temperature and Heat Transfer Mapping for a Waverider at Mach 10 using Fluorescent Paint.* AIAA paper, 94-2484, 1994.

[83] Liu, T.; Campbell, B. T.; Sullivan, J. P.: *Accuracy of Temperature-Sensitive Fluorescent Paint for Heat Transfer Measurements.* AIAA paper, 95-2042, 1995.

[84] Liu, T.; Sullivan J. P.: *Pressure and Temperature Sensitive Paints.* Springer Verlag, Heidelberg, 2005.

[85] Lu, Y.; Bunker, R. S.: *Trench Film Cooling-effect of Trench Downstream Edge and Hole Spacing.* ASME Paper, GT2008-50606, 2008.

[86] Lu, Y.; Dhungel, A.; Ekkad, S. V.; Bunker, R. S.: *Film Cooling Measurements for Cratered Cylindrical Inclined Holes.* ASME Paper, GT2007-27386, 2007.

[87] Lu, Y.; Dhungel, A.; Ekkad, S. V.; Bunker, R. S.: *Film Effectiveness Performance for Coolant Holes Embedded in Various Shallow Trench and CraterDepressions.* ASME Paper, GT2007-27388, 2007.

[88] LU, Y.; NASIR, H.; EKKAD, S. V.: *Film Cooling From a Row of Holes Embedded in Transverse Slots.* ASME Paper, GT2005-68598, 2005.

[89] LUTUM, E.: *Experimentelle Filmkühlungsuntersuchungen zylindrischer und konturierter Bohrungsgeometrien an gekrümmten Oberflächen.* Shaker Verlag, Aachen, 2001.

[90] LUTUM, E.; JOHNSON B. V.: *Influence of the Hole Length-to-Diameter Ratio on Film Cooling With Cylindrical Holes.* ASME J. Turbomach, 121:206–216, 1999.

[91] MAILACH, R.: *Unsteady Flow in Turbomachinery.* TUDpress, Dresden, 2010.

[92] MARGASON, R.: *Fifty Years of Jet in Cross Flow Research.* Proc. AGARD Meeting on Computational and Experimental Assessment of Jet in Cross Flow, 1993.

[93] MCGOVERN, K. T.; LEYLEK, J. H.: *A Detailed Analysis of Film-Cooling Physics: Part II - Compound-Angle Injection With Cylindrical Holes.* ASME J. Turbomach, 122:113–121, 2000.

[94] MENDOZA, D. R.: *An Analysis of CCD Camera Noise and Its Effect on Pressure Sensitive Paint Instrumentation System Signal-To-Noise Ratio.* Experiments in Fluids, 2000.

[95] MENTER, F.: *Two-equation eddy-viscosity turbulence models for engineering applications.* AIAA Journal, 32 (93-2906):1598–1605, 1994.

[96] MHETRAS, S.; HAN, J. C.; RUDOLPH, R.: *Effect of Flow Parameter Variations on Full Coverage Film-Cooling Effectiveness for a Gas Turbine Blade.* ASME J. Turbomach., 134:011004–1 – 011004–10, 2012.

[97] MÜLLER, R., SAUER, H., VOGELER, K., HOEGER, M.: *Influencing the Secondary Losses in Compressor Cascade by a Leading Edge Bulb Modification at the endwall.* ASME Paper GT-2002-30422, 2002.

[98] MOUSSA, Z. M.; TRISCHKA, J. W.; ESKINAZI, S.: *The Near Field in the Mixing of a Round Jet with a Cross Stream.* Journal of Fluid Mechanics, Cambridge Universi'ty Press, 80:49–80, 1977.

[99] NICKLAS, M.: *Film-Cooled Turbine Endwall In A Transonic Flow Field: Part II - Heat Transfer And Film-Cooling Effectiveness.* ASME J. Turbomach, 123:720–729, 2001.

[100] NIEHUIS, R.; LÜCKING, P.; STUBERT, B.: *Experimental and Numerical Study on Basic Phenomena of Secondary Flows in Turbines.* AGARD-CP-469, 1989.

[101] NITSCHE, W.: *Strömungsmeßtechnik.* Springer Verlag, Berlin, 1993.

[102] PEDERSEN, M. C.; ECKERT, E. R. G.; GOLDSTEIN, R. J.: *Film Cooling With Large Density Differences Between Mainstream And Secondary Fluid Measured By Heat Transfer.* ASME J. Heat Transfer, 99:620–627, 1977.

[103] REYNOLDS, O.: *On the dynamical theory of incompressible viscous fluids and the determination of the criterion.* Philosophical Transactions of the Royal Society, 186-

I:123–164, 1895.

[104] Roache, P. J.: *Perspective: A Method for Uniform Reporting of Grid Refinement Studies.* ASME J. Fluids Eng., 116, 1994.

[105] Roache, P. J.: *Quantification of Uncertainty in Computational Fluid Dynamics.* Annual Review of Fluid Mechanics, 29:123–160, 1997.

[106] Roache, P.J.: *Verification and Validation in Computational Science and Engineering.* Hermosa Publishers, Albuquerque, New Mexico, 1998.

[107] Rolls Royce plc.: *The Jet Engine.* 5th edition, Derby, England, 1996.

[108] Sargison, J. E.; Guo S.M.; Oldfield, M. L. G.; Lock, G. D.; Rawlinson, A. J.: *A Converging Slot-Hole Film-Cooling Geometry - Part 1: Low-Speed Flat-Plate Heat Transfer and Loss.* ASME J. Turbomach, 124:453–460, 2002.

[109] Sargison, J. E.; Guo S.M.; Oldfield, M. L. G.; Lock, G. D.; Rawlinson, A. J.: *A Converging Slot-Hole Film-Cooling Geometry - Part 2: Transonic Nozzle Guide Vane Heat Transfer and Loss.* ASME J. Turbomach, 124:460–468, 2002.

[110] Sauer, H., Müller, R., Vogeler, K.: *Reduction of Secondary Flow Losses in Turbine Cascades by Leading Edge Modifications at the Endwall.* ASME J. Turbomach., 123:207–213, 2000.

[111] Sauer, H., Schmidt, R., Vogeler, K.: *Influence of Chord Length and Inlet Boundary Layer on the Secondary Losses of Turbine Blades.* ASME Paper GT-2010-22065, 2010.

[112] Saumweber, C.; Schulz, A.; Wittig, S.: *Free-Stream Turbulence Effects On Film Cooling With Shaped Holes.* ASME Paper, GT-2002-30170, 2002.

[113] Schanze, K. S.; Caroll, B. F.; Korotkevitch, S.; Morris, M. J.: *Temperature dependency of pressure sensitive paints.* AIAA Journal, 35, no.2:306–310, 1997.

[114] Schlichting, H.: *Grenzschicht-Theorie.* Springer Verlag, Berlin Heidelberg, 1996.

[115] Schmidt, D. L.; Bogard, D. G.: *Effects of Freestream Turbulence and Surface Roughness on Film Cooling.* ASME Paper 96-GT-462, 1996.

[116] Schmidt, D. L.; Sen, B.; Bogard, D. G.: *Effects of Surface Roughness on Film Cooling.* ASME Paper 96-GT-299, 1996.

[117] Schmidt, D. L.; Sen, B.; Bogard, D. G.: *Film Cooling with Compound Angle Holes: Adiabatic Effectiveness.* ASME J. Turbomach., 118:807–813, 1996.

[118] Sharma, O.P.; Butler, T.L.: *Predictions of Endwall Losses and Secondary Flows in Axial Flow Turbine Cascades.* ASME J. Turbomach, vol. 109:pp. 229–236, 1987.

[119] Sieverding, C. H.: *Recent Progress in the Understanding of Basic Aspects of Secondary flows in Turbine Blade Passages.* ASME J. Turbomach, 107:248–257, 1985.

[120] Sieverding, C.H.; van den Bosch, P.: *The Use of Coloured Smoke to Visualize*

Secondary Flows in a Turbine-Blade Cascade. Journal of Fluid Mechanics, vol. 134:pp. 85–89, 1983.

[121] Sinha, A. K.; Bogard, D. G.; Crawford, M. E.: *Film-Cooling Effectiveness Downstream of a Single Row of Holes With Variable Density Ratio.* ASME J. Turbomach., 113:442–449, 1991.

[122] Smith, S. H.; Mungal, M. G.: *Mixing, Structure and Scaling of the Jet in Crossflow.* Journal of Fluid Mechanics, Cambridge University Press, 357:83–122, 1998.

[123] Sullivan, J.: *Advanced Measurement Techniques - Temperature and Pressure Sensitive paint.* von Karman Institute for fluid dynamics, Lecture Series, 01, 2001.

[124] Takeishi, K.; Matsuura, M.; Aoki, S.; Sato, T.: *An Experimental Study Of Heat Transfer And Film-Cooling On Low Aspect Ratio Turbine Nozzles.* ASME J. Turbomach, 112:488–496, 1990.

[125] Teekaram, A. J. H.; Forth, C. J. P.; Jones, T. V. : *The Use of Foreign Gas to simulate the Effects of Density Ratios in Film Cooling.* ASME J. Turbomach, 111:57–62, 1989.

[126] VKI: *VKI Lecture Series: Film Cooling Science and Technology for Gas Turbines.* Van Karman Institute for Fluid Dynamics, GT2006-90852, 2005.

[127] Vogel, G.: *Experimental Study on a Heavy Film Cooled Nozzle Guide Vane with Contoured Platforms.* Ph.D. thesis, L'École Polytechnique Fédérale de Lausanne, L'École Polytechnique Fédérale de Lausanne, 2002.

[128] Vogeler, K.: *Zukunftsperspektiven für die Hochtemperatur-Gasturbine.* VGB Powertech PT-10/2011, 2011.

[129] Voigt, M.: *Probabilistische Simulation des strukturmechanischen Verhaltens von Turbinenschaufeln.* Ph.D. thesis, Technische Universität Dresden, Fakultät für Maschinenwesen, Institut für Strömungsmechanik, 2010.

[130] Wagner, G.; Vogel, G.; Chanteloup, D.; Bölcs, A.: *Pressure Sensitive Paint (PSP) and Transient Liquid Crystal Technique (TLC) for Measurements of Film Cooling Performances.* 16th Symposium on Measuring Techniques in Transonic and Supersonic Flow in Cascades and Turbomachines, 2002.

[131] Walters, D. K.; Leylek, J. H.: *A Detailed Analysis of Film-Cooling Physics: Part I - Streamwise Injection With Cylindrical Holes.* ASME J. Turbomach, 122:102–112, 2000.

[132] Wang, H.P.; Olson, S.J.; Goldstein, R.J.; Eckert, E.R.G.: *Flow Visualization in a Linear Turbine Cascade of High Performance Turbine Blades.* ASME J. Turbomach., vol. 119:pp. 1–8, 1997.

[133] Wedlake, E. T.; Brooks, A. J.; Haragasma, S. P.: *Aerodynamic And Heat Transfer Measurements On A Transonic Nozzle Guide Vane.* ASME Paper 88-GT-10,

1988.

[134] WEISS, A.: *Der Einflus der Profilgeometrie auf die Entwicklung der Sekundärströmungen in Turbinengittern.* Ph.D. thesis, Universität der Bundeswehr München, UniBw München, Fakultät für Luft- und Raumfahrttechnik, 1993.

[135] WEIGAND, B.; CRAWFORD, M. E.; LUTUM, E.: *A Computational and Experimental Investigation of the Effect of Surface Roughness on Film Cooling.* In: ISROMAC-7 Intern. Symposium on Transport Phenomena and Dynamics of Rotating Machinery, 1998.

[136] WENDT, F.W.: *Computational Fluid Dynamics.* Springer Verlag, New York, 1992.

[137] WICK, A.: *Ein numerisches Verfahren zur Stroemungssimulation in zeitveraenderlichen Gebieten mit integriertem Modul zur Gitternachfuehrung.* Ph.D. thesis, Technische Universität Berlin, Fakultät für Verkehrs- und Maschinensysteme, 2003.

[138] WILCOX, D. C.: *Multiscale Model for Turbulent Flows.* AIAA 24th Aerospace Sciences Meeting, American Institute of Aeronautics and Astronautics, 1986.

[139] WILFERT, G.; WOLFF, S.: *Influence of Internal Flow on Film Cooling Effectiveness.* ASME J. Turbomach, 122:327–333, 2000.

[140] WRIGTH, L. M.; MCCLAIN, S. T.; CLEMENSON, M. D.: *Effect of Density Ratio on Flat Plate Film Cooling with Shaped Holes using PSP.* ASME J. Turbomach., 134:011004–1 – 011004–10, 2011.

[141] YUAN, L. L.; STREET, R. L.; FERZIGER, J. H.: *Large-eddy Simulations of a Round Jet in Crossflow.* Journal of Fluid Mechanics, Cambridge University Press, 379:71–104, 1999.

[142] YUEN, C. H. N.: *Measurement of local heat transfer coefficient and effectiveness in film cooling geometries.* Ph.D. thesis, University of Higher Education, University of London, Imperial College of Science, Technology and Medicine, 2000.

[143] YUEN, C. H. N.; MARTINEZ-BOTAS, R. F.; WHITELAW, J. H.: *Film Cooling Effectiveness Downstream of Compound and Fan-Shaped Holes.* ASME 2001-GT-0131, 2001.

[144] ZHANG, L. J.; JAISWAL, R. S.: *Turbine Nozzle Endwall Film Cooling Study Using Pressure-Sensitive Paint.* ASME J. Turbomach, 123:730–738, 2001.

A Appendix

A.1 Uncertainty Estimation

In general the uncertainty analysis that is performed in this section uses the method of error propagation as described by Bevington and Robinson [9] considering the possible sources of error. Maximum error limits and the uncertainty is conducted for experimental results as the static wall pressures, blowing rates, discharge coefficients, total pressure losses and flow field variables as well as for the surface temperature and subsequently the film effectiveness.

Each measured quantity is subject to errors which can be divided into bias errors, random errors and general errors. Assuming a quantity y is determined by a measured input parameter x that is afflicted with errors, the influence of x to the derived quantity y can be described by using a first order Taylor-series expansion for sufficiently small values of the absolute error Δx.

$$y(x+\Delta x) = y(x) + \frac{1}{1!}\frac{dy}{dx}\,\Delta x \tag{A.1}$$

$$\Delta y = y(x+\Delta x) - y(x) = \frac{dy}{dx}\,\Delta x \tag{A.2}$$

If y is a function of several independent input variables x_1, ..., x_i that are subject to errors, the Taylor-series expansion is truncated after the linear term. The total differential with its partial derivatives is then written with respect to the input variables as

$$\Delta y = \left|\frac{\partial y}{\partial x_1}\right| \Delta x_1 + \left|\frac{\partial y}{\partial x_2}\right| \Delta x_2 + ... + \left|\frac{\partial y}{\partial x_i}\right| \Delta x_i. \tag{A.3}$$

Herein Δy represents the maximum error of y as function of the input variables (x_1, ... x_i) where their error sources will propagate into the quantity y. Following the Gaussian method of error propagation, each input variable x_i is considered not to be correlated and contains a number of defective quantities ν_j. For that reason the input variable is replaced by its mean value along with its uncertainty u_i which is determined by the sample variance as

$$u_i^2 = var(x_i) = \frac{1}{(N-1)} \sum_{j=1}^{N} (\nu_j - \overline{\nu})^2. \tag{A.4}$$

The uncertainty of quantity y is then obtained by quadratic summation of the individual uncertainties u_i (random variation of the measurement variables x_i) as

$$u_y = \sqrt{\left(\frac{\partial y}{\partial x_1} u_1\right)^2 + \left(\frac{\partial y}{\partial x_2} u_2\right)^2 + ... + \left(\frac{\partial y}{\partial x_i} u_i\right)^2}. \tag{A.5}$$

Further it can be rearranged to an equivalent form given by

$$\frac{var(y)}{y^2} = \sum_{i=1}^{N} \frac{var(x_i)}{x_i^2} \frac{x_i^2}{y^2} \left(\frac{\partial y}{\partial x_i}\right)^2. \tag{A.6}$$

With the definition of the sensitivity coefficient $S_i = \frac{x_i}{y} \frac{\partial y}{\partial x_i}$ it yields

$$\frac{var(y)}{y^2} = \sum_{i=1}^{N} S_i^2 \frac{var(x_i)}{x_i^2}. \tag{A.7}$$

A.1.1 Static Wall Pressures

Static wall pressures on the endwall as well as on the airfoil surface are measured by a rotary pressure system (Scanivalve SSS-48C). It contains a differential pressure sensor with a pressure range of 1 PSID and a measurement accuracy of 0.06 % full scale. Applying the method of error propagation to the definition of the pressure coefficient C_{p2} as defined in equation (5.4) it delivers the maximum uncertainty ΔC_{p2}.

Hereby it needs to be considered that the velocity head $p_{t2} - p_2$ composes of two separately measured values. The surface pressure p_x and the total pressure p_{t2} are obtained by differential pressure transducers. These sensors measure the pressure differences versus the ambient pressure as $\Delta p_{x,M} = p_x$ - p_{amb} and $\Delta p_{t2,M} = p_{t2} - p_{amb}$. For the present case, the static pressure p_2 equals to the ambient pressure since the cascade expands to ambient conditions. Therefore the surface pressure can be rewritten as $\Delta p_{x,M} = p_x - p_2$ and the velocity head is reduced to $p_{t2} - p_2 = \Delta p_{t2,M}$. Herewith it follows for the uncertainty in C_{p2}

$$\Delta C_{p2}^2 = \left(\Delta(\Delta p_{x,M}) \left|\frac{\partial C_{p2}}{\partial \Delta p_x}\right|\right)^2 + \left(\Delta(\Delta p_{t2,M}) \left|\frac{\partial C_{p2}}{\partial \Delta p_{t2,M}}\right|\right)^2 \tag{A.8}$$

which can be rewritten to

$$\Delta C_{p2} = \sqrt{\left(\Delta(\Delta p_{x,M}) \left|\frac{1}{\Delta p_{t2,M}}\right|\right)^2 + \left(\Delta(\Delta p_{t2,M}) \left|\frac{-\Delta p_{x,M}}{\Delta p_{t2,M}^2}\right|\right)^2}. \tag{A.9}$$

For the estimation of uncertainties, the absolute measurement error of the surface pressure with $\pm 4.1\,Pa$ is taken into account. Furthermore, the velocity head $p_{t2} - p_2$ equals the measured total pressure difference Δp_{t2} at the cascade exit. It is obtained via a Setra D239 pressure sensor with a pressure range of 0 .. $3.8\,kPa$. Considering the measurement accuracy of this sensor with 0.10 % full scale it equals an absolute pressure uncertainty of $\pm 3.8\,Pa$. The ambient pressure is obtained by a Setra B270 transducer with a pressure range of 80 ... $110\,kPa$. The sensor offers a higher accuracy of 0.03 % full scale which equals to an error of $\pm 9\,Pa$ for the ambient pressure reading.

With that, the uncertainty in the pressure coefficient for the airfoil pressure distributions ranges from ± 0.0019 to ± 0.0029 which results into an averaged relative uncertainty of approximately 1.2 %. In regions of very low pressure coefficients, which occur near the trailing edge or locally on the pressure side, the relative uncertainty of C_{p2} may exceed 8 %.

Similar uncertainty estimates are observed for the endwall pressure distributions. The absolute uncertainty of the pressure coefficient ranges from ± 0.0019 to ± 0.0042 yielding also an averaged value of about 1.2 %. Endwall pressure coefficients significantly decrease toward zero at the exit region of the cascade. In this endwall area, the absolute uncertainty is quite low and the relative uncertainty for C_{p2} may exceed 15 - 20 % due to very low surface pressure differences conducted at often less than two Pascal.

When the uncertainty of the isentropic Mach number is considered, its definition given in equation (5.5) is differentiated to apply the method of error propagation. The resulting total differential then contains the measurement error of the surface pressure which equals to $\pm 4.1\,Pa$ since the scanivalve system was used and the error of the total pressure of the incoming flow p_{t0}. This quantity is obtained from the velocity head obtained by the Prandtl-probe $\Delta p_{Pr} = p_{t0} - p_0$, the separately measured static pressure difference Δp_0 and the ambient pressure as $p_{t0} = \Delta p_{Pr} + \Delta p_0 + p_{amb}$.

The measurement error for p_{t0} is estimated to $\pm 16.6\,Pa$. The uncertainty of the isentropic Mach number shows deviations within 0.00060 - 0.00594 for a range of $0.02 < Ma_{is} < 0.20$. Furthermore the uncertainty distribution was averaged to obtain a global uncertainty estimate for Ma_{is} with ± 0.00163 which equals 3.7 %. It is smaller for higher isentropic Mach numbers as 0.3 % for higher Ma_{is} up to 0.2 and exceeds 30 % for Ma_{is} lower than 0.02.

A.1.2 Total Pressure Losses

The measurement uncertainty of the secondary loss coefficient is influenced by various individual error sources. Based on the definition of the spatial total pressure loss coefficient as given in equation (5.9) the considered terms can be rearranged to represent the measured pressure differences as

$$\zeta_{2,l} = \frac{p_{t0} - p_{t2}}{p_{t0} - p_2} = \frac{\Delta p_t}{(p_{t0} - p_{t2}) + (p_{t2} - p_2)} = \frac{\Delta p_t}{\Delta p_t + (p_{t2} - p_2)} \tag{A.10}$$

Applying the method of error propagation leads to the uncertainty as

$$\Delta\zeta_2^2 = \left(\Delta(\Delta p_t)\left|\frac{\partial\zeta_2}{\partial(\Delta p_t)}\right|\right)^2 + \left(\Delta(p_{t2})\left|\frac{\partial\zeta_2}{\partial p_{t2}}\right|\right)^2 + \left(\Delta(p_2)\left|\frac{\partial\zeta_2}{\partial p_2}\right|\right)^2 \tag{A.11}$$

which can be rewritten in terms of the measured quantities as

$$\Delta\zeta_2^2 = \left(\Delta(\Delta p_t)\left|\frac{\Delta p_{t2}}{(\Delta p_t + (p_{t2} - p_2))^2}\right|\right)^2 + \left(\Delta(\Delta p_{t2})\left|\frac{-\Delta p_t}{(\Delta p_t + (p_{t2} - p_2))^2}\right|\right)^2 + \ldots \quad (A.12)$$

$$\ldots + \left(\Delta(\Delta p_2)\left|\frac{\Delta p_t}{(\Delta p_t + (p_{t2} - p_2))^2}\right|\right)^2 . \quad (A.13)$$

whereas index 0 refers to upstream of cascade (location of Prandtl-probe) and 2 to the measurement plane at cascade exit. Table A.1 summarizes all important informations of the pressure transducers used during the loss measurements along with their uncertainties. It is obvious that the measured total pressure difference $p_{t0} - p_{t2}$ has a major impact to the uncertainty of the total pressure loss coefficient. It composes of the total pressure measured at the Pitot-probe p_{t2} and the total pressure p_{t0} for the free-stream far upstream of the cascade which is determined by a Prandtl-probe. The total pressure difference is obtained by a sensor cascade connecting three pressure transducers with different pressure ranges. Dependent on the pressure level of

transducer	measurement value	pressure range	accuracy			
			factory		calibrated	
			[%FS]	[Pa]	[%FS]	[Pa]
D239	$p_{t0} - p_{t,i}$	0 ... 632 Pa	0.14	0.9	0.1	0.6
		0 ... 1247 Pa	0.14	1.7	0.1	1.2
		0 ... 3736 Pa	0.14	5.2	0.1	3.8
D239	$p_{t2} - p_2$	0 ... 3736 Pa	0.14	5.2	0.1	3.8
D239	$p_{t0} - p_0$	0 ... 3736 Pa	0.14	5.2	0.1	3.8
D239	$p_0 - p_2$	-1868 ... 1868 Pa	0.14	5.2	0.1	3.8
B270	p_{amb}	80 ... 110 kPa	0.05	15	0.03	9

Table A.1: Summary of used pressure transducers for the total pressure loss measurements

the obtained value, transducers are switched to the optimum sensor range to minimize the measurement uncertainty. For this Setra pressure transducers D239 are used which provide pressure ranges of 0 ... $632\,Pa$, 0 ... $1.2\,kPa$ and 0 ... $3.8\,kPa$. All sensors are calibrated and offer a low measurement uncertainty of $0.10\,\%$ fullscale. This equals to absolute measurement accuracies for the pressure reading of $\pm\,0.6$, $\pm\,1.2$ and $\pm\,3.8\,Pa$ respectively. The loss coefficient is furthermore affected by the isentropic dynamic pressure $p_{t0} - p_2$. It is composed of the directly measured pressure differences $\Delta p_t = p_{t0} - p_{t2}$ and the total pressure of the pitot probe $\Delta p_{t2} = p_{t2} - p_2$. It therefore yields an increased error of $\pm\,12.8\,Pa$ combining the measurement accuracy for both pressure transducers.

Furthermore, errors are introduced by the Keithley digital multimeter during the read-out and the digitizing of the voltage signals. The used DMM 2700 of the Integra series a voltage range of 0 ... 10 V with a resolution of $\pm 10\mu V$. The accuracy for voltage reading is given with ± 0.002 % reading + ± 0.0005 %. Considering a typical voltage range of a Setra D239 with 0 ... 5 V the absolute error due to voltage reading is $\pm 0.000125\, V$ or equally expressed as relative error of ± 0.0025 %. For the smallest pressure sensor the absolute error due to read-out and digitizing is then estimated with $\pm 0.016\, Pa$. It is therefore significantly smaller by the factor of 40 and can be neglected for further considerations.

As a matter of fact, the influence of the total pressure difference Δp_t on the uncertainty of the spatial loss coefficient is larger by a factor of three or higher compared to the ambient pressure p_2 and absolute total pressure p_{t2}. But the influence that these values add to the uncertainty of the loss coefficient is of controversy character. To identify the trends, table A.2 summarizes the results of the uncertainty estimation carried out at two specific measurement locations. It displays the primary variables that are used to determine the secondary loss coefficient along with their uncertainty.

measurement uncertainty for local loss coefficient ζ_2	value	unit	uncertainty		
			location 1	location 2	location 3
	$\Delta(p_{t0} - p_{t2})$	[Pa]	0.6	0.6	0.6
	$\Delta(p_{t2} - p_2)$	[Pa]	22.4	22.4	22.4
	ΔT_1	[K]	0.216	0.216	0.216
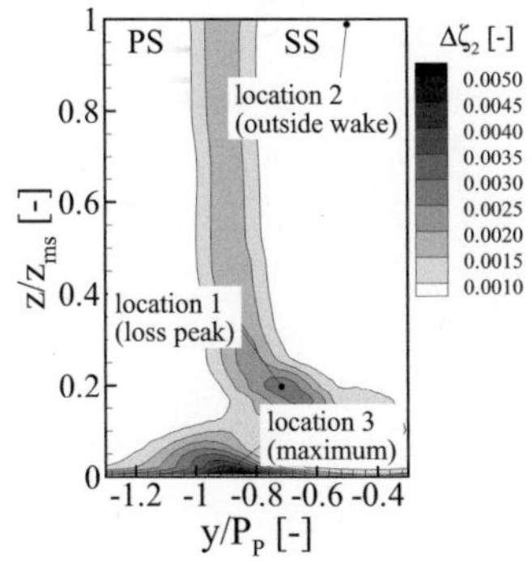	Δu_2	[m/s]	0.252	0.221	0.320
	$\Delta u_{2,x}$	[m/s]	0.095	0.072	0.069
	$\Delta u_{2,y}$	[m/s]	0.232	0.207	0.312
	$\Delta \rho\, u_{2,x}$	[kg/m²s]	0.287	0.255	0.366
	$\Delta \beta_2$	[°]	0.1	0.1	0.1
	$\Delta \zeta_{1,l}$	[-]	0.00113	0.00113	0.00113
	$\Delta \zeta_{2,l}$	[-]	0.00168	0.00056	0.00375
	$\Delta \overline{\zeta}_1$	[-]	0.00114	0.00114	0.00114
	$\Delta \overline{\zeta}_2$	[-]	0.00169	0.00057	0.00376
	$\Delta \zeta_{ew}$	[-]	0.00317	0.00080	0.00664
	$\Delta \zeta^n_{ew,0}$	[-]	0.01203	0.00264	0.02417

Table A.2: Summary of estimated measurement uncertainties for the total pressure loss measurements obtained for the test case of $\beta_{1,ms} = 94.8°$ at several measurement positions

In regions of low total pressure losses, the magnitude of Δp_t decreases significantly while the obtained absolute total pressure increases. With that and bearing in mind that the measurement accuracy of the pressor sensors is related to their full scale range, small or zero total pressure difference has significant influence on the uncertainty of the loss coefficient $\Delta\zeta_2$. Although the absolute uncertainty levels are observed to be quite small in such regions, the relative uncertainty may increase up to 25 - 60 %.

Considering the total pressure losses obtained at location 2, the total pressure difference is found to be significantly low at 3 Pa since it is outside the wake at midspan. The absolute total pressure measures $p_{t2} = 2211\,Pa$ which equals a spatial loss coefficient of $\zeta_2 = 0.00104$. Hence, the uncertainty of the loss coefficient is $\Delta\zeta_2 = 0.00056$ which gives a relative uncertainty of 56 %. When the measured Δp_t increases, the total pressure of the pitot probe significantly decreases and the loss coefficient rises. Therefore its influence on the uncertainty of the loss coefficient is reduced but it is still larger than the influence of the absolute total pressure. Due to that, higher magnitudes of the loss coefficient results into higher absolute uncertainties for ζ_2 but the uncertainty decreased to less 1 % with respect to the obtained loss coefficient.

As an example, the uncertainty of the local loss coefficient is found to be 0.00168 at location 1 where the loss peak represents the trailing edge vortex sheet. The total pressure difference is found to have risen up to 516 Pa along with a decreased absolute total pressure of $p_{t2} = 1700\,Pa$. Herewith it equals to a relative error of ± 0.72 % with regard to the obtained loss coefficient of $\zeta_2 = 0.233$. Furthermore additional error sources would be introduced by applying the area- and mass averaging method as well as the reduction and normalization of the spatial loss coefficient to obtain the net endwall loss. While taking these influences into account, the variation of the uncertainty for the secondary loss coefficient is typically of one order of magnitude between ± 0.002 and ± 0.026 depending on the measured total pressure difference. The relative uncertainty of the net endwall loss is then estimated to approximately ± 2 ... 2.5 % for measurement regions with high total pressure loss. In the free-stream region where only small or no pressure loss is obtained, the uncertainty may exceed ± 10 %.

A.1.3 Film Effectiveness

Applying the method of error propagation to the definition of the film effectiveness gives the uncertainty as

$$\Delta\eta_f^2 = \left(\Delta T_m \left|\frac{\partial \eta_f}{\partial T_m}\right|\right)^2 + \left(\Delta T_f \left|\frac{\partial \eta_f}{\partial T_f}\right|\right)^2 + \left(\Delta T_c \left|\frac{\partial \eta_f}{\partial T_c}\right|\right)^2 \tag{A.14}$$

which can be transformed into

$$\Delta\eta_f = \sqrt{\left(\Delta T_m \left|\frac{(T_f - T_c)}{(T_m - T_c)^2}\right|\right)^2 + \left(\Delta T_f \left|\frac{(T_c - T_m)}{(T_m - T_c)^2}\right|\right)^2 + \left(\Delta T_c \left|\frac{(T_m - T_f)}{(T_m - T_c)^2}\right|\right)^2} \tag{A.15}$$

wherein the uncertainties for the main flow and coolant temperature are primarily influenced by the measurement errors introduced by the thermocouples and the digital multimeter. In contrast, the uncertainty for the surface temperature acquired with TSP is attributed to the used optical system and the selected luminescent paint formulation. It contains several error sources that need to be addressed in order to establish quantitative measurements. Therefore Liu and Sullivan [84] conducted a detailed uncertainty analysis for intensity-based TSP as well as PSP-measurement techniques since many of the error sources are the same for both systems. The elemental error sources regarding the TSP-technique are discussed by Cattafesta and Liu [25], Liu et al. [83] and Liu et al. [81]. The accuracy of the luminescent paint technique is dependent on various error sources which can be divided into three major groups as displayed in Fig. A.1.

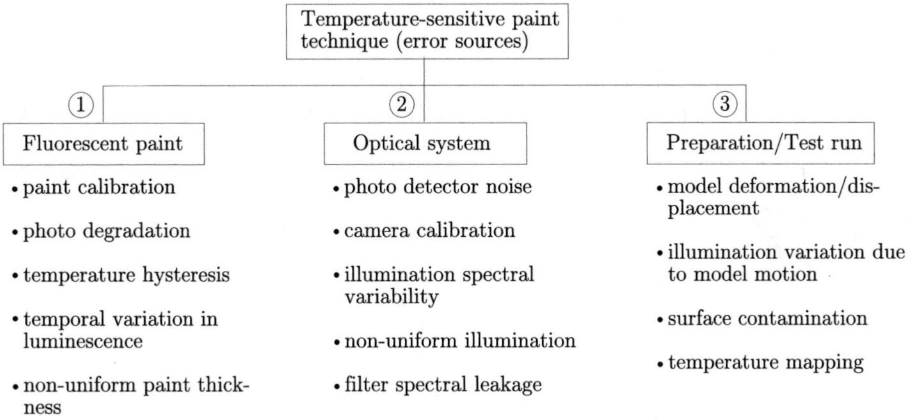

Figure A.1: Possible error sources of intensity-based TSP-measurement technique

The first group contains errors as paint calibration, photo degradation and temperature hysteresis which are intrinsically related to the chemical and physical properties of the paint. Errors in the second group are mostly related to the measurement system. They include uncertainties induced by noise, non-linearity and non-uniformity of the photo detector as well as spectral variability and filter leakage of the illumination system. Additional errors (third group) can be introduced during the preparation of the experimental setup and during the test runs which can be caused by deformation and displacement of the model geometry. An uneven paint thickness or non-uniform illumination may also contribute to the uncertainty but can significantly be reduced by taking a ratio of the acquired test image to a reference image.

An uncertainty analysis formalism was stated by Liu and Sullivan [84] wherein functional relationships were developed for the system components to describe the elemental error sources and their sensitivity coefficients in the error propagation. According to this formalism for TSP-measurements, the total uncertainty of temperature T is described by the error propagation equation

$$\frac{var(T)}{(T - T_{ref})^2} = \sum_{i=1}^{N} S_i^2 \frac{var(\xi_i)}{\xi_i^2} \tag{A.16}$$

assuming that the cross correlation coefficients between the variables vanish. The variables ξ_i ($i = 1 \ldots N$) denote a set of parameters describing the elemental errors influencing the total uncertainty of the temperature. These parameters are associated with photo degradation, sedimentation, filter leakage and spectral variability, model deformation as well as paint calibration and the luminescent intensity measurement via a photo detector. The sensitivity coefficient S_i is defined as

$$S_i = \frac{\xi_i}{(T - T_{ref})} \frac{\partial T}{\partial \xi_i}. \tag{A.17}$$

In order to conduct the uncertainty analysis for the TSP-technique, a brief discussion will follow pointing out the major contributing error sources. Table A.3 shows a summary of all considered elemental errors along with their physical origin and quantitative estimates of the variances.

variable	physical origin	sensitivity coefficient S_i	elemental variance $\frac{var(\xi_i)}{\xi_i^2}$	estimated relative error [%]
$D_{q0}(\Delta t)$	temporal variation of illumination		$\left(\frac{\Delta I}{I}\right)^2$	0.1
$D_t(\Delta t)$	photo degradation surface contamination		$\left(\frac{\Delta D_t}{D_t}\right)^2$	0.17
$\Pi_f/\Pi_{f,ref}$	illumination spectral variability & filter leakage		$\left(\frac{\Delta \Pi_f}{\Pi_f}\right)^2$	0.03
$D_x(\Delta x)$	model deformation/ model displacement		$\left(\frac{\Delta x}{x}\right)^2$	0.07
V_{ref}, V	photo detector noise reference / signal image		$V_{ref} G \hbar \nu B_d$ $V G \hbar \nu B_d$	0.13
ΔT_{cal}	paint calibration		$\left(\frac{\Delta T_{cal}}{T_{cal}}\right)^2$	0.15 - 0.23

Table A.3: Estimation of elemental errors for TSP-measurements

The uncertainty associated with the photo detector (e.g. CCD-camera, photodiode or photomultiplier tube) is related to various noise sources such as photon shot noise, image sensor noise and read-out noise. The photon shot noise is attributed to the inherent uncertainty of electromagnetic radiation which is described best by a statistical distribution with an uncertainty. It is an unavoidable quantum effect that occurs when a photo detector collects a number of photons from a steady light source and can be statistically described by a Poisson distribution (Cattafesta et al. [25]). The photon shot noise is characterized by the signal-to-noise ratio (SNR) to quantify the quality of light signal detection in relation to the total noise of the optical system. It is defined by the ratio of generated charge carriers (signal electrons) to the sum of noise signals as

$$\frac{S}{N} = \frac{n_{signal}}{n_{noise}} = \frac{\frac{\Phi}{\hbar\nu}\, t\, A\, \eta}{\sqrt{\left(\sqrt{\frac{\Phi}{\hbar\nu}\, t\, A\, \eta}\right)^2 + n_{CCD}^2 + n_{readout}^2}} \tag{A.18}$$

with Φ ... the light power [W/m^2], A ... the pixel area [m^2], $\hbar\nu$ - the photon energy [Ws] with $\hbar$ - Planck's constant and ν the light frequency defined by $\nu = c/\lambda$ (c - speed of light, λ_c - wavelength). Further quantities are involved as η - the quantum efficiency and t - the exposure time [s]. The number of signal electrons are dependent on the illumination intensity of the incoming photons. The total noise signal is composed of the photon noise, read-out noise and the noise floor produced by the CCD image sensor. Since all three noise sources are assumed not to be correlated, the total number of noise electrons is given by the variance $var(n_{noise})$. Fig. A.2 displays the relation between the generated charge carriers (signal electrons) to the impinging intensity of the illumination light or light power onto the CCD-sensor. If there is a strong light signal, the number of signal electrons that are generated would be much larger than the noise electrons produced by the image sensor. In that case the photon noise is dominant giving a simplified signal-to-noise ratio which can be estimated by $\sqrt{A\,\eta}$.

Considering low-light conditions, the noise that is generated by the image sensor would dominate over the light signal resulting into a poor SNR. Herein the signal-to-noise ratio for the used PCO 1600 camera is found to be SNR $= 46\,dB$ at its full-well capacity of 40000 e$^-$. According to Crafton [28] the shot noise is reduced by the square root of number of images acquired and the total number of photons collected during measurement yielding $n_{shot} \propto 1/\sqrt{(N \cdot n_{e^-})}$. Assuming a photon-noise limited case with an averaged pixel of the CCD-sensor being saturated at 70 % of its full-well capacity, the acquisition of a number of 20 images yields an estimated shot noise of 0.13 %.

The read-out noise is added during the conversion of the charge carriers into a voltage signal and the subsequent data processing as well as analog-digital conversion. It is caused by the amplifier processing chain and the A/D-conversion. The read-out noise for the used CCD-camera is given with 10 e$^-$ rms at $10\,MHz$. Different noise influences can occur on a CCD-image sensor which include dark current noise, pattern noise, transfer noise (additional or lost charge carriers during sensor's shift between the registers) or noise from external sources. In low-light situations with long exposure times the dark noise gains more influence on the image. This effect thermally induces charge carriers that are misinterpreted as voltage signals. This noise level can be reduced by thermoelectric cooling of the sensor. For the PCO 1600 it reduces the dark current noise down to 0.01 e$^-$ by cooling down to temperatures of -20 $°C$.

Considering a photon-shot noise limited case without any model deformation, the limiting temperature resolution of a photo detector-paint system can be estimated by a method suggested by Liu and Sullivan [84]. It takes into account the paint sensitivity and the photon shot noise generated by the camera neglecting all other influencing parameters. Furthermore, the

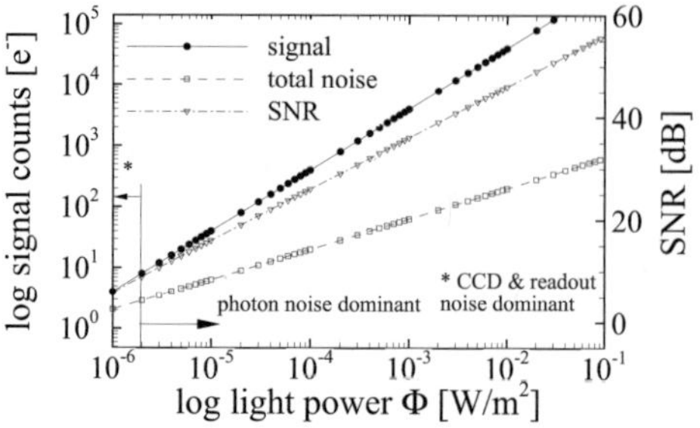

Figure A.2: Light signal, total noise and signal-to-noise ratio (SNR) versus incoming light intensity in double logarithmic scale, detector specifications: $\lambda_c = 580\,nm$, $A = 54.8\,\mu\text{m}^2$, $t = 50\,ms$, $\eta = 0.5$ (at $580\,nm$), CCD read-out noise = 10 e^- rms (at $10\,MHz$), full-well capacity = 40000 e^-

photon shot noise is significantly reduced by averaging a number of images. Due to this, the temperature resolution increases by the factor $N^{\frac{1}{2}}$.

$$\Delta T_{\min} = \frac{K_t}{\sqrt{N\, n_{e,\max}}}\left[1 + \exp\left(\frac{T - T_{ref}}{K_t}\right)\right]^{\frac{1}{2}}. \tag{A.19}$$

Herein $n_{e,max}$ is the full well capacity of the CCD-sensor which characterizes the maximum amount of charge for each pixel before saturation occurs. It is dependent on the pixel size and the number of images N that are averaged. The UniTSP-01 from Innssi Inc. has an averaged sensitivity factor K_t over its temperature range of $56.9\,K$. The minimum resolvable temperature difference for the used TSP-system has been estimated to $\Delta T = 0.4\,K$ for a single frame of image. Enhancement in temperature resolution is found by averaging a larger number of images as displayed in Fig. A.3. It decreases from $0.40\,K$ for a single image down to a temperature resolution of $\Delta \text{T} = 0.08\,K$ at $T = 293\,K$ for $N = 20$. The temperature resolution slightly degrades about $0.02\,K$ ($T = 343\,K$) with increased temperature due to reduced change in luminescent intensity.

A non-uniform illumination and uneven paint thickness can be neglected by rationing test and reference image. Temporal variation of luminescent intensity mainly results either from photo degradation or sedimentation of dust or oil droplets onto the painted surface. Photo degradation occurs when the paint is exposed to the strong illumination light that can cause a significant decrease in luminescent intensity over time. The process depends on the illumination intensity, exposure time and the paint itself. In order to estimate the amount of photo degradation for these experiments, the used TSP-paint was constantly illuminated over a time period of 180 minutes at 50 % and maximum illumination intensity. The temperature of the paint was held constant within $\pm 0.2\,K$ over the entire exposure time. The experiments reveal a degradation

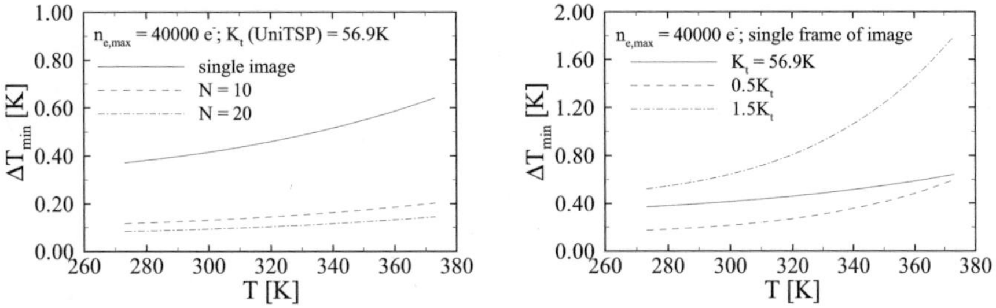

Figure A.3: Minimum temperature resolution of UniTSP: camera system PCO1600, 14-bit dynamic A/D range, full well capacity 40000e$^-$, quantum efficiency 65 dB at 550 nm

effect at 50 % illumination intensity that is far less than 1 % over the entire exposure time. It results in a photo degradation effect of 0.3 % per hour. To minimize errors caused by photo degradation, the exposure time should therefore be as short as possible. According to Cattafesta et al. [25] the signal-to-noise ratio for a severely degraded paint decreases as the signal becomes considerably weaker.

Sedimentation is caused by dirt and oils drop which are transported in the flow and accumulate on the painted surface. Over the measurement time the sediments shield the paint from the illumination light and mask the paint luminescence from the detector. Installed filters at the wind tunnel inlet avoid any visible sedimentation during the experiments. This effect can therefore be estimated to less than 0.5 % per hour. To further reduce the effects of photo degradation and sedimentation, the measurement time (which is time period between the wind tunnel starts until the final wind-on images are acquired) are shortened as possible. Up to five data sets of 20 images at 30 frames per second were taken for each image series. Since the LED-lamps are covered between the wind-off and the wind-on conditions, the exposition of the paint to the illumination source is limited to the data acquisition time for each image series. The time required to record 5 image sets is estimated to take three minutes. This results in photo degradation of 0.018 %. A time span of 15 minutes after coolant injection must be taken into account to allow the system to come to a thermal equilibrium. Thus, the painted surface is exposed to the flow so that sedimentation can take place over about 18 minutes. This results into a sedimentation effect of 0.155 %.

Spectral leakage may occur when the luminescence signal is contaminated by wavelengths of the illumination light. It is often caused by improperly selected filter combinations for camera and illumination system. Thereby small portions of excitation and ambient light are allowed to reach the detector to negatively affect the luminescent signal. This error can be minimized by selecting appropriate filters to separate the wavelength spectra of the excitation light source and the luminescent signal. For these experiments short-pass filtering starts at 500 nm and the long-pass filters transmit at a wavelength of 540 nm. That leads to a resulting gap of about 40 nm to distinguish between both wavelength spectra. With that the filter leakage is estimated of 0.01 %. The spectral output of LED-lamps is quite stable but requires a warm-up phase of several minutes. The variability in the LED-spectrum is less than 0.1 % after 30 minutes of warm-up and can be neglected.

One of the major sources of uncertainty for measuring the wall temperature is related to the calibration of TSP-paint whereas the measured intensity I/I_{ref} is rated to the temperature obtained by embedded surface thermocouples. To obtain a functional dependency for the calibration, a parametric relationship as fourth-degree polynomial is applied to the calibration data by

$$T_{cal} = a_1 \frac{I}{I_{ref}}^4 + a_2 \frac{I}{I_{ref}}^3 + a_3 \frac{I}{I_{ref}}^2 + a_4 \frac{I}{I_{ref}} + a_5 \tag{A.20}$$

wherein the coefficients a_i are determined by a least-squares fit. The corresponding uncertainty in temperature is

$$var(T_{cal}) = \sum \left(\Delta a_i \frac{\partial T_{cal}}{\partial a_i} \right)^2 + \left(\Delta \frac{I}{I_{ref}} \frac{\partial T_{cal}}{\partial \frac{I}{I_{ref}}} \right)^2 . \tag{A.21}$$

Expecting the uncertainties associated with the fitting coefficients to be negligibly small, the error attributed to the measured intensity ratio dominates the uncertainty of the calibration. With that, equation (A.21) is reduced to

$$\Delta T_{cal} = \left[\Delta \frac{I}{I_{ref}} \frac{\partial T_{cal}}{\partial \frac{I}{I_{ref}}} \right] . \tag{A.22}$$

To obtain the TSP-calibration curve, a representative sample of 10000 values is obtained within a region of 100 pixel x 100 pixel on the calibration plate. The data is extracted from the intensity images for each calibration temperature. Hereby care must be taken in selecting the size of samples. Small samples would lead to high statistical uncertainty and dominate the results of the calibration. One way to avoid this influence is to increase the size of the samples which will decrease the variance and reduces the length of the 95%-confidence level (Voigt [129]). The in-house tool ProSi-Post, established by Voigt [129], was used to determine the statistical parameters as mean value and standard deviation for each calibration point.

The histogram of the luminescent intensity ratio, displayed in Fig. A.4 for the calibration temperature of $T = 306\,K$, exhibits a Gaussian distribution. Similar behavior is observed for all other calibration points as well. Considering the case of $T = 306\,K$, the standard deviation of the intensity ratio is obtained to $\sigma = 0.0035$ with its mean value at $I/I_{ref} = 0.7732$. These quantities are obtained for a confidence level of 95.4 % giving the (1-α) confidence interval of $\pm\,0.0075$. Considering that the standard deviation is defined as $\sigma = \sqrt{var(x_i)}$ the uncertainty of the luminescent intensity ratio can be described by u $(I/I_{ref}) = 2\sigma(I/I_{ref})$ with regard to the

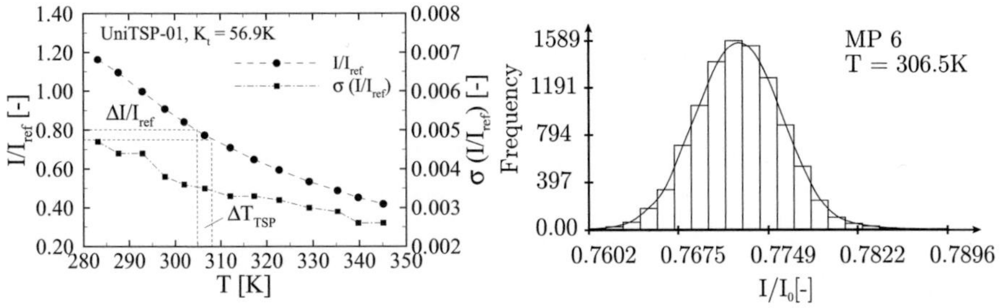

Figure A.4: Intensity ratio I/I_{ref} with its standard deviation σ (left) and histogram of calibration point at $T = 306\,K$ (right)

confidence level of 95.4 %. According to equation (A.22), the uncertainty in paint calibration is then obtained to $\pm\,0.52\,K$.

Since the paint sensitivity varies over the TSP-temperature range ($T = 290\ ...\ 340\,K$), as illustrated in Fig. A.5, the calibration uncertainty ΔT_{cal} does change between $\pm\,0.46\,K$ to $\pm\,0.79\,K$. The sensitivity of the paint is quite remarkably reduced toward the limits of the temperature range. Exceeding those boundaries would result into enhanced sensitivity factors K_t and increased measurement uncertainties. Falling below the lower temperature limit means that the paint emits reduced intensity change. This can be further enhanced by condensation and icing issues due to the humidity of the ambient air. Also above the upper temperature limit, the luminescence reaction of the probing molecules steadily decreases. With that it might drop below the minimum intensity change that is required to resolve a specific change in surface temperature. Hence it is necessary to arrange the temperature conditions of the experiment so that the surface temperature is kept within those limits.

The total uncertainty for the temperature measured with the TSP-technique ΔT_f is then computed according to equation (A.16) by summing all considered elemental error sources as listed in table A.3. Due to the variation of the paint sensitivity which results into varying standard deviation of the luminescent intensity ratio and hence the calibration temperature, the total uncertainty of surface temperature varies between $\pm\,0.74\,K$ and $\pm\,1.0\,K$ (see also Fig. A.5).

To finally obtain the uncertainty for the film effectiveness, the individual uncertainties for all temperatures entering the three-temperature problem, as given in equation (A.15) must be considered. The uncertainty for the main flow ΔT_m and coolant static temperatures ΔT_c are primarily influenced by the measurement accuracy of the thermocouple sensors. Hence, the sensors are pre-calibrated versus a high-precision resistance thermometer PT100 offering a calibrated measurement accuracy of $\pm\,0.01\,K$. With that, the absolute measurement accuracy is estimated with $\pm\,0.2\,K$. As only total temperatures can be obtained with these thermocouples,

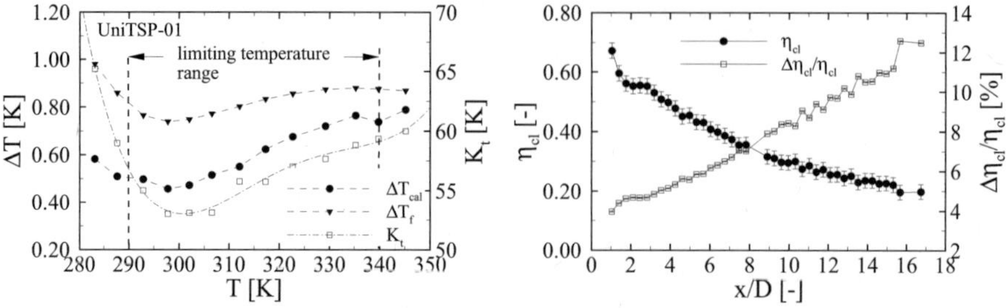

Figure A.5: Temperature uncertainty of the paint calibration ΔT_{cal} and total uncertainty for the surface temperature ΔT_f measured via TSP with sensitivity factor K_t as well as the resulting measurement uncertainty of film effectiveness for a selected test case with flat plate film-cooling: cylindrical holes $M = 0.5$, $D_R = 1.37$

fluid	value	sensor range	sensor accuracy		pressure accuracy
		[Pa]	%FS	[Pa]	[Pa]
main flow	Δp_{tm}	± 1900	0.10	3.8	12.8
	Δp_m	0 ... 7500	0.10	7.5	16.5
coolant	Δp_{tc}	0 ... 7500	0.10	7.5	16.5
	Δp_c	0 ... 7500	0.10	7.5	16.5
	p_{bar}	80 ... 110 kPa	0.03	33	33

Table A.4: Measurement accuracy for total and static pressures during the flat plate film-cooling experiments

the isentropic relationship is used to calculate the static temperatures. This adds additional error sources to the total uncertainty of the film effectiveness induced by the ratio of the total to the static pressure. The measurement accuracy of the pressure differences as well as for the barometric pressure is summarized in Table A.4.

Applying the method of error propagation to the isentropic relationship delivers the uncertainty of the static temperatures for the coolant $\Delta T_c = \pm 0.23\,K$ and the main flow $\Delta T_m = \pm 0.22\,K$. Accordingly, the measurement uncertainty of the film effectiveness is estimated for a test case with cylindrical holes on a flat plate. A small blowing rate of $M = 0.5$ combined with an elevated density ratio of $D_R = 1.37$ was selected. The results of the uncertainty estimation on the centerline film effectiveness are displayed in Fig. A.5. Hereby, the absolute uncertainty is observed to be $\Delta\eta = \pm 0.0267$ closely downstream of the holes which decreases slightly in streamwise direction to $\Delta\eta = \pm 0.0247$. However, the relative measurement uncertainty of the film effectiveness steadily increases in streamwise direction. It is observed at 4 % for η= 0.6 directly downstream of the hole exit. The uncertainty of the film effectiveness increases up to about 10 % in the far field of the holes showing film effectiveness values of $\eta = 0.2$. Uncertainties for very low film effectiveness magnitudes may be higher due to the low temperature gradient. This is inherently connected to small luminescence reaction of the molecules and therefore reduced intensity change.

A.2 Summary of Pressure Transducers used in the Coolant Supply System

No.	location	sensor	pressure value	pressure ranges [KPa]	signal [V]	accuracy (FS) [%]	accuracy (FS) [Pa]
1	air supply	B270	p_{bar}	80 - 110	0-5	0.03	33
2	air supply	PMP4100	p_{Kessel}	0 - 700	0-10	0.04	280
3	air supply	PMP4100	p_{TRZ}	0 - 700	0-10	0.04	280
4	wind tunnel	D239	$p_{t0} - p_0$	0 - 3.8	0-5	0.10	3.8
5	wind tunnel	D239	$p_{t0} - p_{t2}$	0 - 3.8	0-5	0.10	3.8
6	wind tunnel	D239	p_{t2}	0 - 3.8	0-5	0.10	3.8
7	wind tunnel	D239	$p_{w,1}$	± 1.9	± 2.5	0.10	3.8
8	coolant supply	PMP4100	$p_{0.1}$	0 - 700	0-10	0.04	280
9	coolant supply	PMP4100	$p_{0.2}$	0 - 700	0-10	0.04	280
10	coolant supply	PMP4100	$p_{1.1}$	0 - 700	0-10	0.04	280
11	coolant supply	PMP4100	$p_{1.2}$	0 - 700	0-10	0.04	280
12	plenum cavity A	PMP4100	$p_{A.1}$	0 - 70	0-10	0.04	28
13	plenum cavity A	PMP4100	$p_{A.2}$	0 - 35	0-10	0.04	14
14	plenum cavity B	PMP4100	$p_{B.1}$	0 - 100	0-10	0.04	40
15	plenum cavity B	PMP4100	$p_{B.2}$	0 - 35	0-10	0.04	14
16	plenum cavity C	PMP4100	p_C	0 - 100	0-10	0.04	40
17	plenum cavity D	PMP4100	$p_{D.1}$	0 - 100	0-10	0.04	40
18	plenum cavity D	PMP4100	$p_{D.2}$	0 - 35	0-10	0.04	14
19	plenum cavity E	PMP4100	p_E	0 - 70	0-10	0.04	28

Table A.5: Parameters of used pressure transducers in the cascade coolant supply

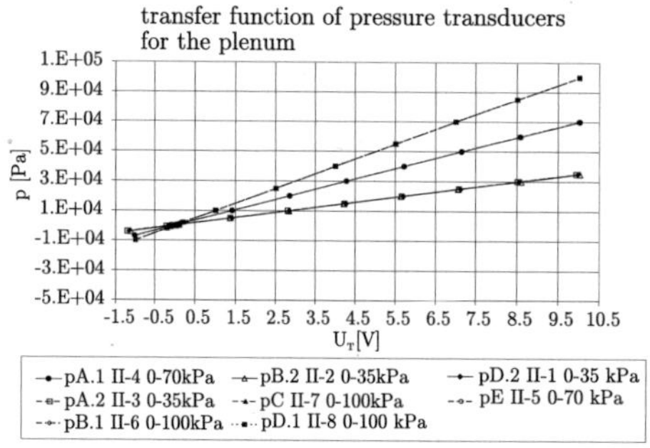

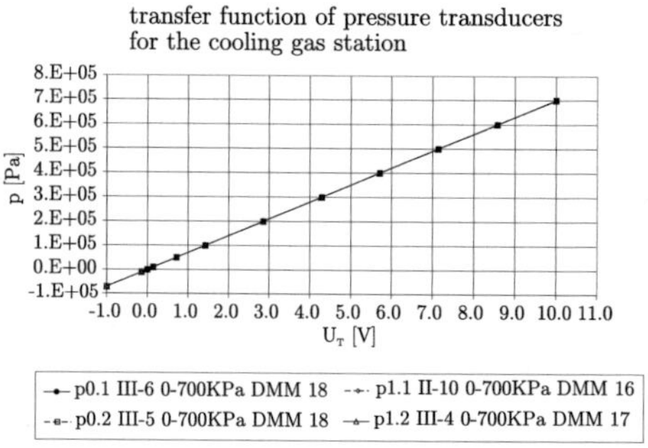

Figure A.6: Transducer voltage signal U_T over calibrated pressure range

A.3 Heat Transfer Coefficients

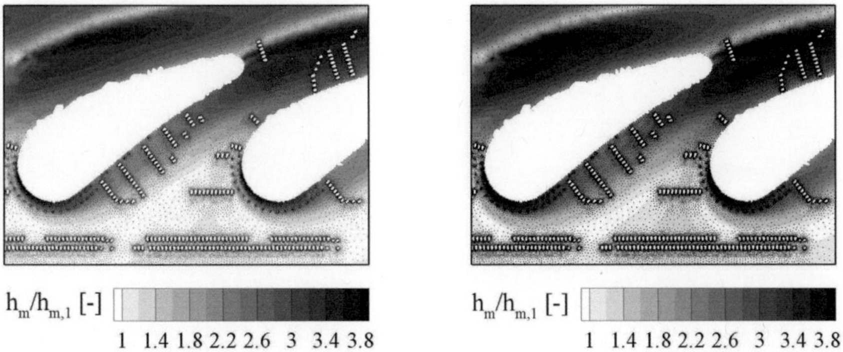

(a) $\beta_{1,ms} = 75.0°$ (b) $\beta_{1,ms} = 49.4°$

Figure A.7: Heat transfer coefficient at cascade endwall for both elevated loading cases; CFD-results mapped onto the nodes of the endwall FE-mesh

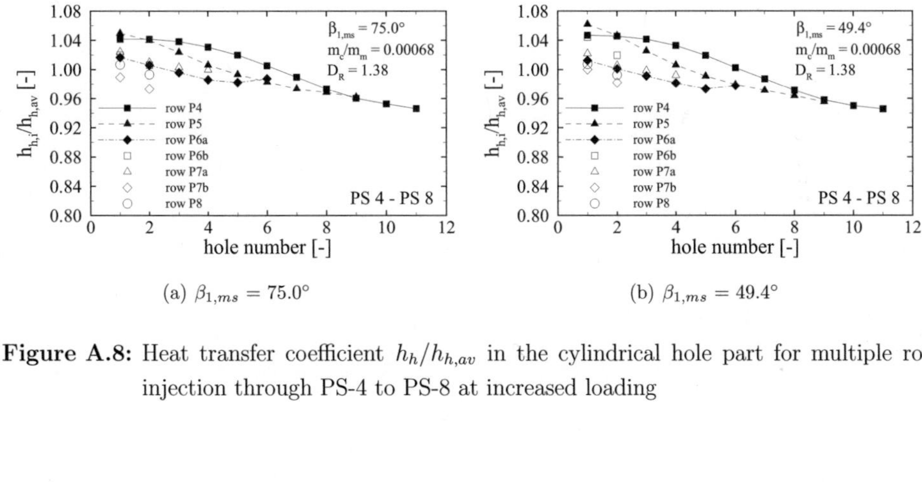

(a) $\beta_{1,ms} = 75.0°$ (b) $\beta_{1,ms} = 49.4°$

Figure A.8: Heat transfer coefficient $h_h/h_{h,av}$ in the cylindrical hole part for multiple row injection through PS-4 to PS-8 at increased loading

A.4 Fillet Effect on Cascade Flow Field

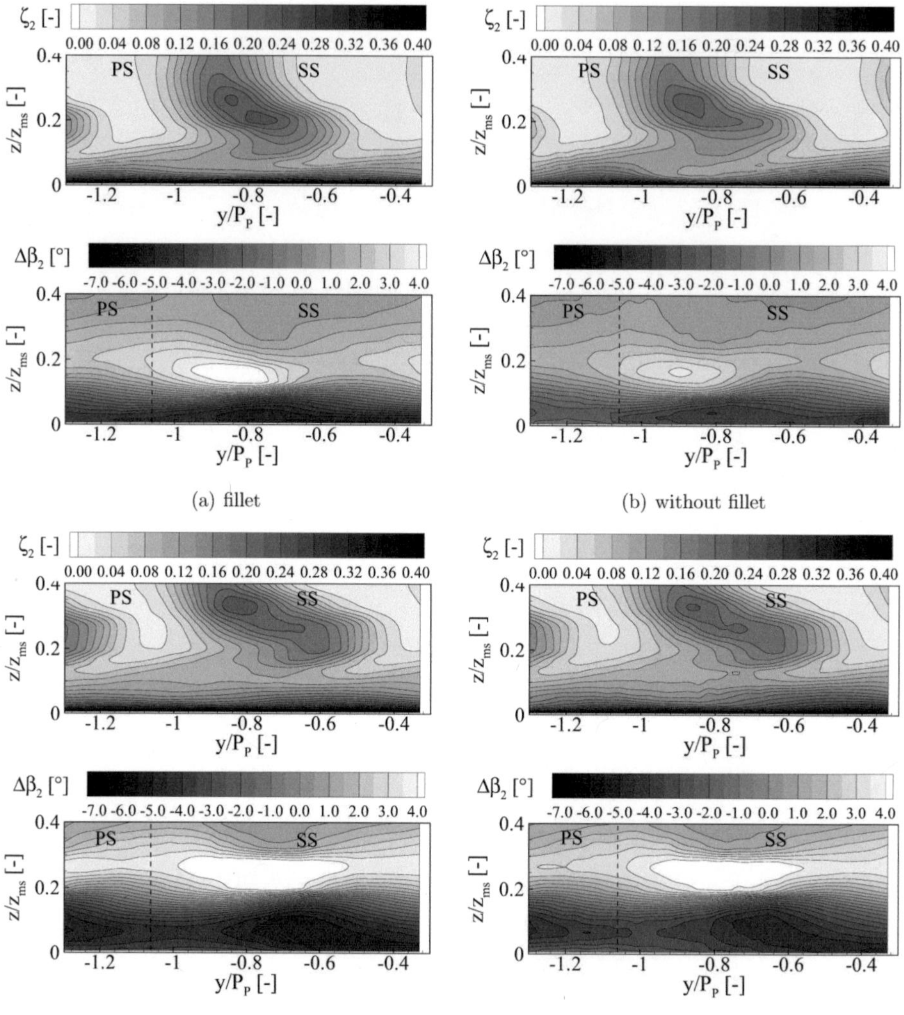

(a) fillet

(b) without fillet

(c) fillet

(d) without fillet

Figure A.9: Effect of the fillet geometry on pressure loss distribution and flow angle deviation, $\beta_{1,ms} = 75.0°$ (a-b) and $\beta_{1,ms} = 49.4°$ (c-d), $P/L_P = 0.7$, $s/L_P = 1.03$

A.5 5-Hole Probe Measurements at Cascade Outlet

A.5.1 Total Pressure Loss and Secondary Flow Vectors

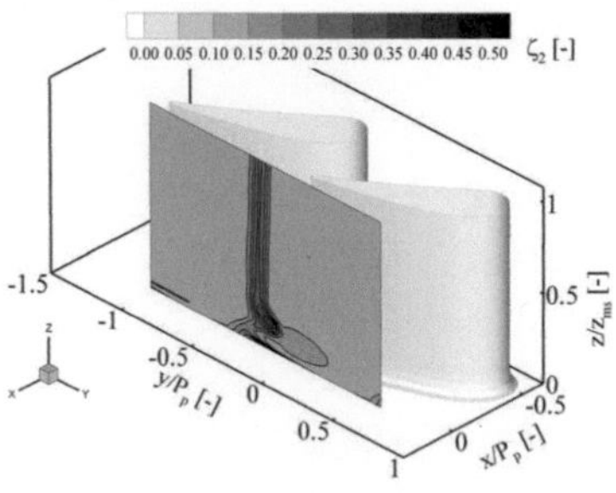

(a) s/L$_P$ = 0.34

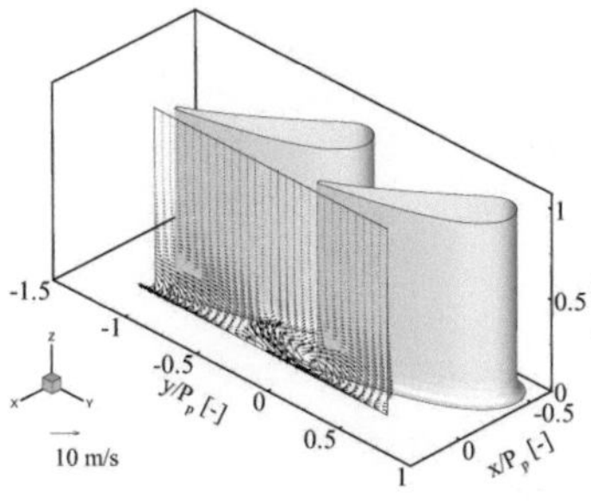

(b) s/L$_P$ = 0.34

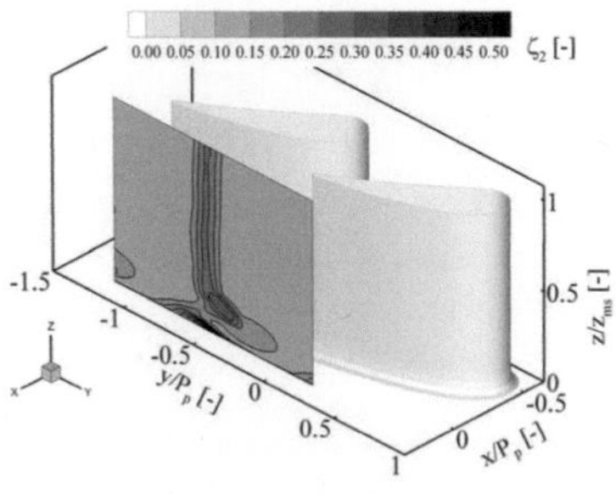

(c) s/L$_P$ = 0.70

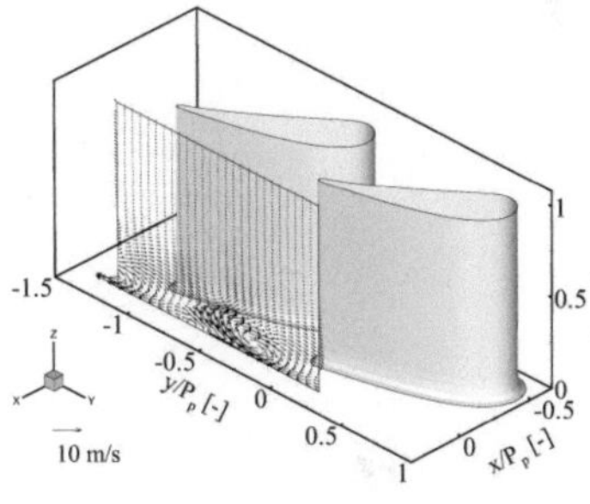

(d) s/L$_P$ = 0.70

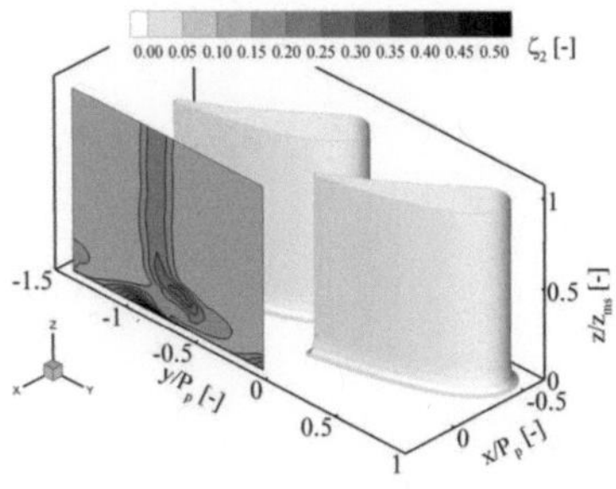

(e) s/L$_P$ = 1.03

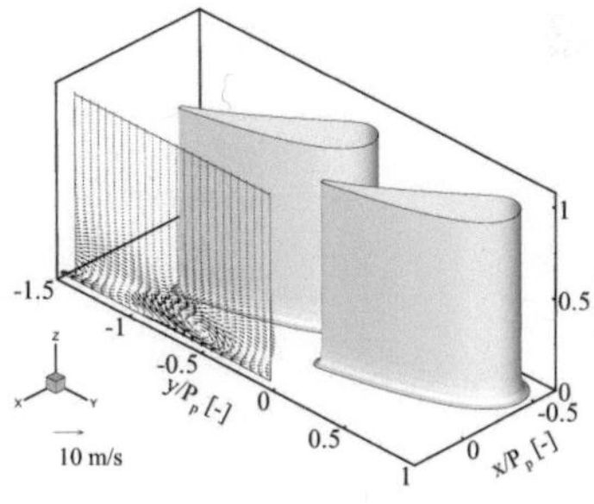

(f) s/L$_P$ = 1.03

Figure A.10: Total pressure loss coefficient (a,c,e) and secondary flow vectors (b,d,f) with varying streamwise distances s/L_P, loading case $\beta_{1,ms} = 94.8°$

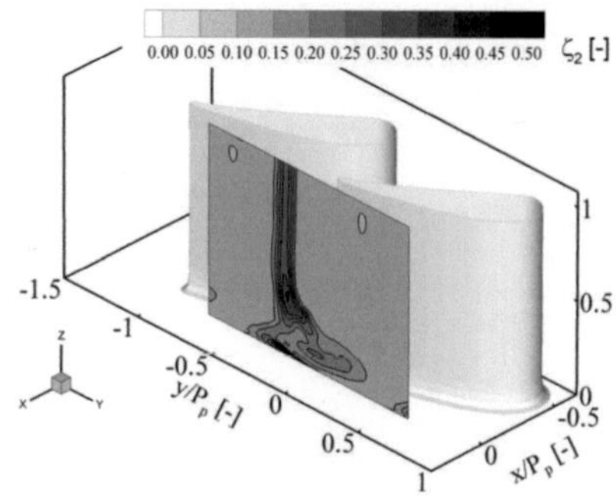

(a) s/L$_P$ = 0.34

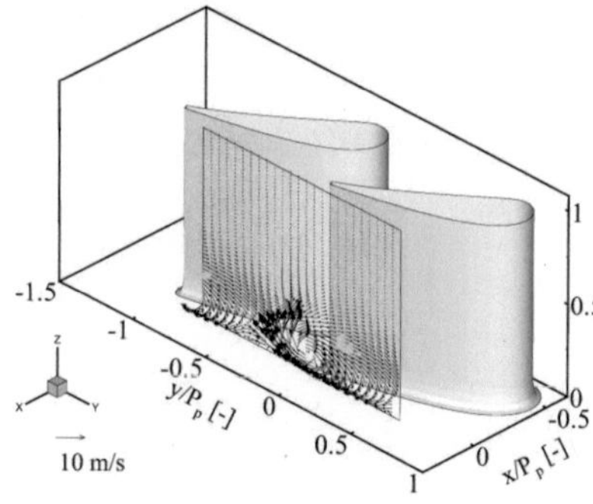

(b) s/L$_P$ = 0.34

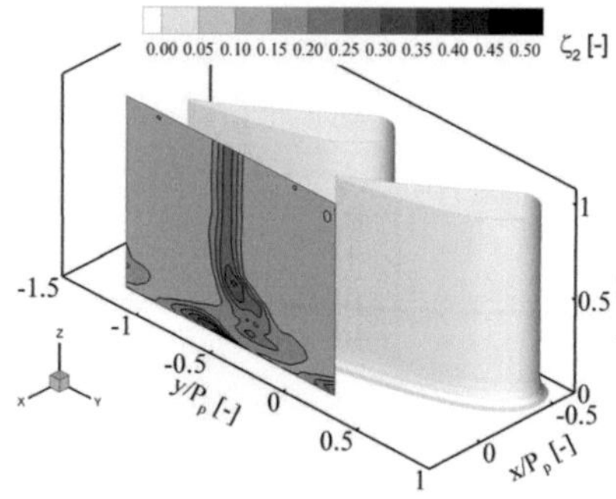

(c) s/L$_P$ = 0.70

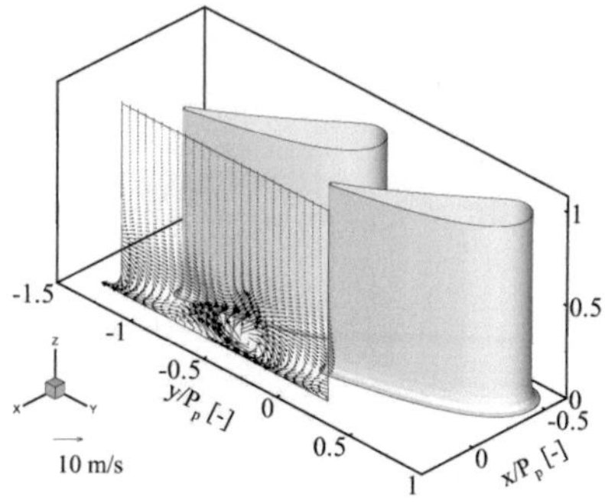

(d) s/L$_P$ = 0.70

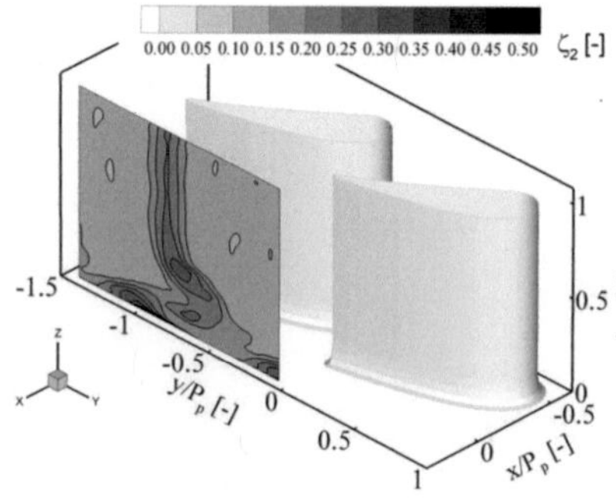

(e) s/L$_P$ = 1.03

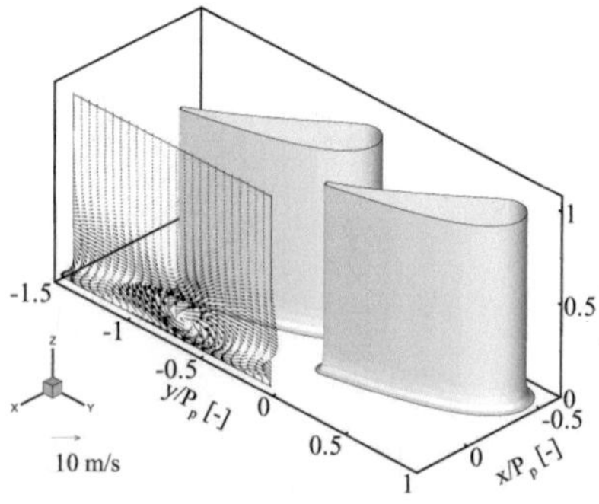

(f) s/L$_P$ = 1.03

Figure A.11: Total pressure loss coefficient (a,c,e) and secondary flow vectors (b,d,f) with varying streamwise distances s/L_P, loading case $\beta_{1,ms} = 75.0°$

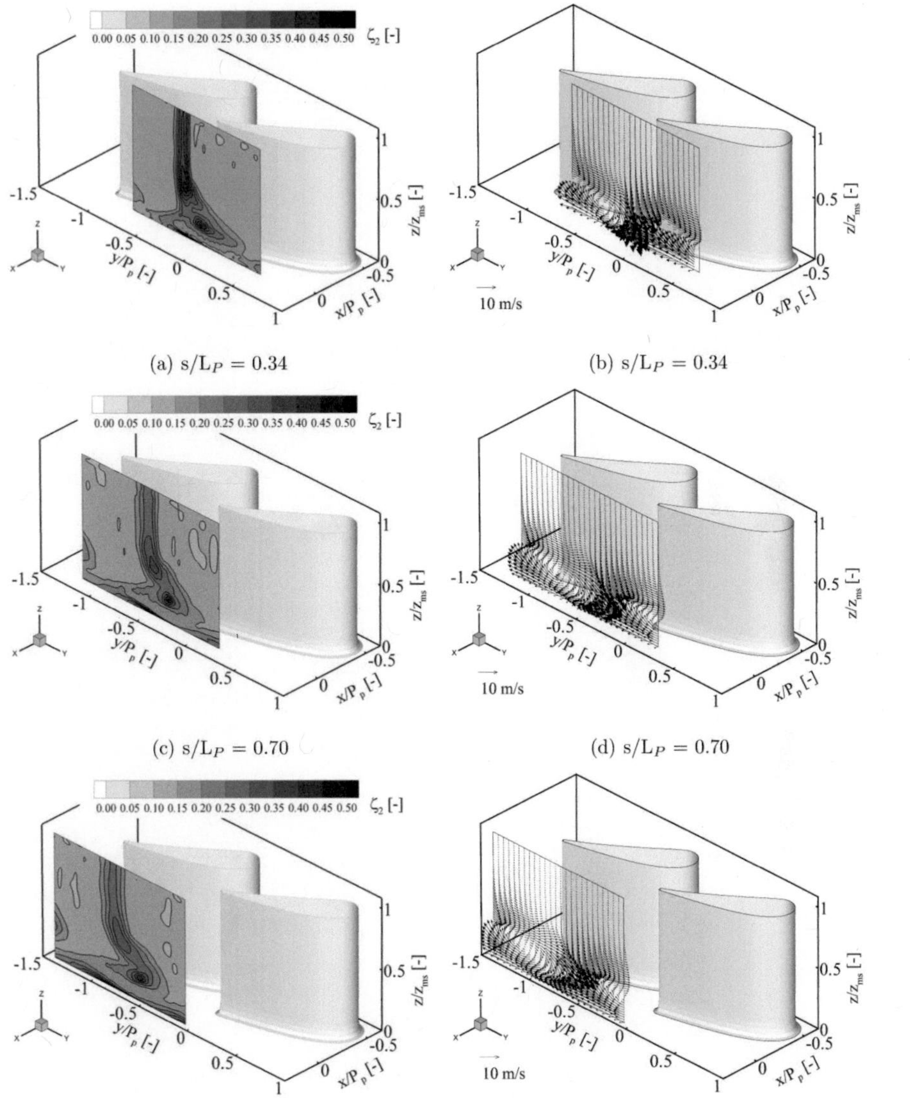

(a) s/L$_P$ = 0.34 (b) s/L$_P$ = 0.34

(c) s/L$_P$ = 0.70 (d) s/L$_P$ = 0.70

(e) s/L$_P$ = 1.03 (f) s/L$_P$ = 1.03

Figure A.12: Total pressure loss coefficient (a,c,e) and secondary flow vectors (b,d,f) with varying streamwise distances s/L_P, loading case $\beta_{1,ms} = 49.8°$

A.5.2 Streamwise Vorticity (SVO)

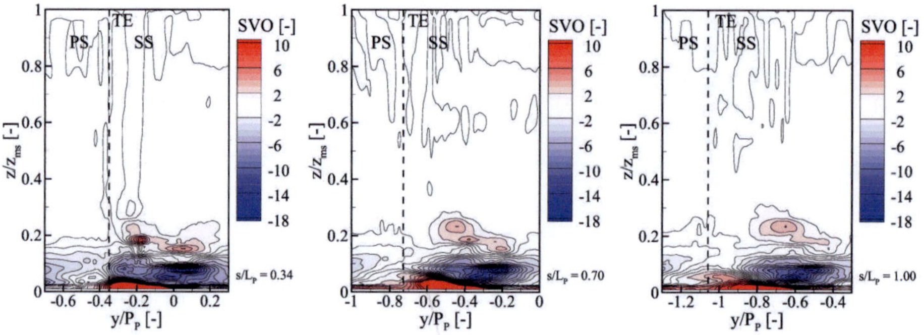

(a) $\beta_{1,ms}$=95.4°

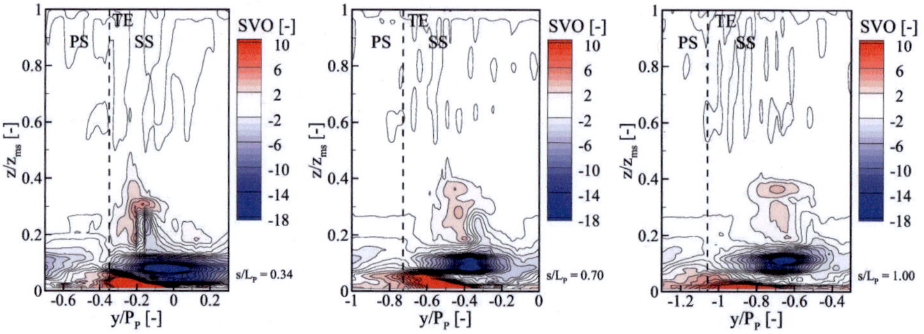

(b) $\beta_{1,ms}$=75.0°

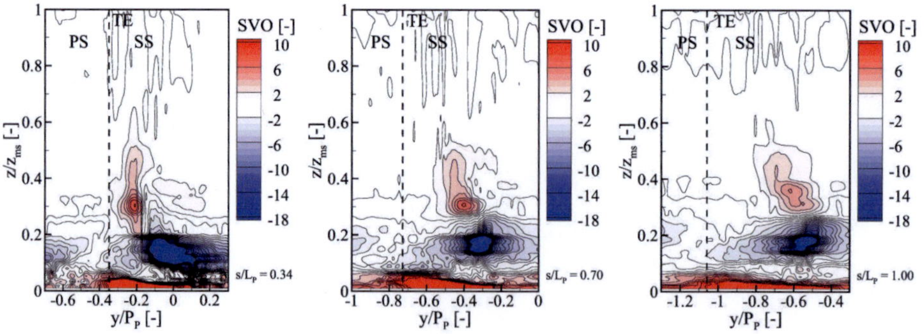

(c) $\beta_{1,ms}$=49.8°

Figure A.13: Streamwise Vorticity (SVO) at cascade exit for varying streamwise distance $s/L_P = 0.34$, 0.70 and 1.03 (from left to right) and variation of cascade loading